Schriftenreihe der Professur für Molekulare Lebensmitteltechnologie

Band 8

Institut für Ernährungs- und Lebensmittelwissenschaften
Professur für Molekulare Lebensmitteltechnologie

Role of glutathione and *Botrytis cinerea* laccase activities in wine quality

Dissertation

zur Erlangung des Grades

Doktor der Naturwissenschaften (Dr. rer. nat.)

der Mathematisch-Naturwissenschaftlichen Fakultät
der Rheinischen Friedrich-Wilhelms-Universität Bonn

von

Sabrina Zimdars

aus

Marburg

Bonn 2020

Angefertigt mit Genehmigung der Mathematisch-Naturwissenschaftlichen Fakultät der Rheinischen Friedrich-Wilhelms-Universität Bonn

1. Gutachter: Prof. Dr. Andreas Schieber
IEL - Molekulare Lebensmitteltechnologie, Universität Bonn

2. Gutachter: Prof. Dr. André Lipski
IEL - Lebensmittelmikrobiologie und -hygiene, Universität Bonn

Tag der Promotion: 11.05.2020

Erscheinungsjahr: 2020

Bibliografische Information der Deutschen Nationalbibliothek
Die Deutsche Nationalbibliothek verzeichnet diese Publikation in der Deutschen Nationalbibliografie; detaillierte bibliographische Daten sind im Internet über http://dnb.d-nb.de abrufbar.
1. Aufl. - Göttingen: Cuvillier, 2020
Zugl.: Bonn, Univ., Diss., 2020

Nonnenstieg 8, 37075 Göttingen
Telefon: 0551-54724-0
Telefax: 0551-54724-21
www.cuvillier.de

1. Auflage, 2020
Gedruckt auf umweltfreundlichem, säurefreiem Papier aus nachhaltiger Forstwirtschaft.
Titelbild (links) gedruckt mt der Genehmigung des Bildeigentümers, Martin Ladach.

ISBN 978-3-7369-7254-4
eISBN 978-3-7369-6254-5

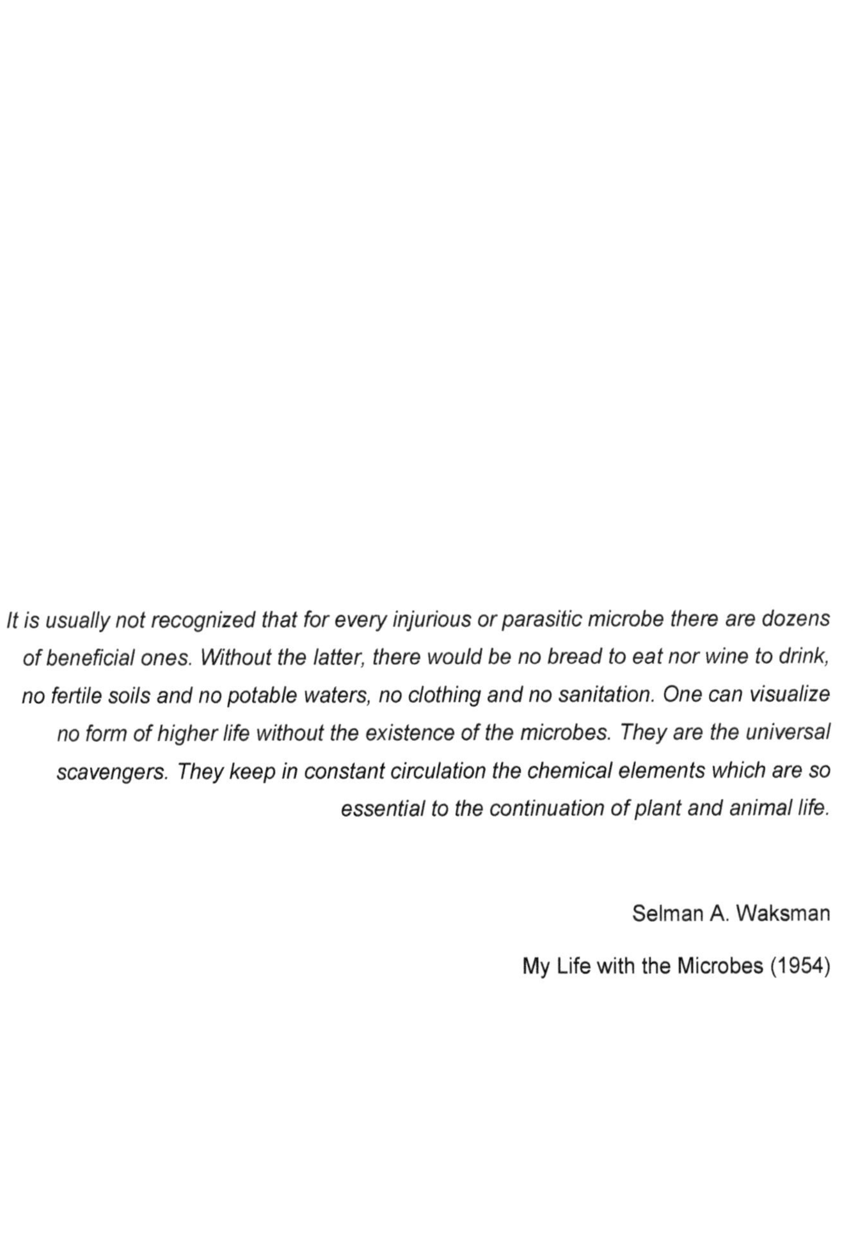

It is usually not recognized that for every injurious or parasitic microbe there are dozens of beneficial ones. Without the latter, there would be no bread to eat nor wine to drink, no fertile soils and no potable waters, no clothing and no sanitation. One can visualize no form of higher life without the existence of the microbes. They are the universal scavengers. They keep in constant circulation the chemical elements which are so essential to the continuation of plant and animal life.

Selman A. Waksman

My Life with the Microbes (1954)

Table of contents

Preliminary remarks

List of abbreviations

ANOVA	analysis of variance
AAS	atomic absorption spectroscopy
CoA	coenzyme A
cys	cysteine
DAD	diode array detector
EDTA	ethylenediaminetetraacetic acid
eV	electron volt
FCM	flow cytometry
FID	flame ionization detector
FSC	forward scatter
g	g-force
GC	gas chromatography
glu	glutamate
gly	glycine
GSH	glutathione
GSSG	glutathione disulfide

HPLC	high-performance liquid chromatography
kDa	kilodalton
m/z	mass-to-charge ratio
MS	mass spectrometry
MV	microvinification
n	number of samples
n.d.	not detected
PPO	polyphenoloxidase
RI	retention index
SSC	side scatter
U	enzyme activity units
(U)HPLC	(ultra) high-performance liquid chromatography
UPLC	ultra-performance liquid chromatography
UV	ultraviolet
Vis	visible
v/v	volume fraction of two liquids

List of publications

Zimdars, S.; Schrage L.; Schieber, A.; Weber, F. Influence of glutathione on yeast fermentation efficiency under copper stress. *Journal of Agricultural and Food Chemistry*, **2019**, 67, 10913–10920.

Zimdars, S.; Hitschler J.; Schieber A.; Weber, F. Oxidation of wine polyphenols by secretomes of wild *Botrytis cinerea* strains from white and red grape varieties and determination of their specific laccase activity. *Journal of Agricultural and Food Chemistry*, **2017**, 65, 10582–10590.

Conferences

Zimdars, S.; Schieber, A.; Weber, F. Impact of glutathione on yeast vitality during fermentation of *Botrytis*-infected musts. 4th Edition of the International Conference Series on Wine Active (WAC) compounds in Beaune, France, 29–31 March, **2017**, *Book of Abstracts*, 158. [Poster]

Declaration of contribution as co-author

The following co-authors contributed to the papers presented in Chapter 2 and 3:

Prof. Dr. Andreas Schieber contributed to the publications as the supervisor of this thesis and proofread all manuscripts.

Dr. Fabian Weber advised on the design of experimental work and supported the interpretation and publication of the results. As the corresponding author, he proofread the manuscripts and handled the formal aspects of the publications.

Dr. Stephan Sommer assisted in the interpretation of the results of the paper in Chapter 2.

Julia Hitschler supported the selection of a suitable method for the cultivation of *Botrytis cinerea* isolates, the determination of the laccase activity, and substrate oxidation tests. She further contributed to the subsequent UHPLC analysis.

Lukas Schrage assisted in the conduction of microvinifications and selected a suitable method for the ultrasound disruption of yeast cells and determination of alcohol dehydrogenase activity. He further supported the analysis of yeast vitality and fermentation parameters with flow cytometry, HPLC, and GC.

Chapter 1

General Introduction

1 Fungal infections in the vineyard

Winemaking has a long history representing one of the oldest food processing techniques performed by humans since ancient civilizations. The earliest records of wine production were found between 5400 and 5000 BC in Iran at the Hajji Firuz Tepe site (Pérez-Torrado et al., 2018). Over the centuries, wine was part of the human culture concerning dietary and socio-religious aspects (Soleas et al., 1997), and it has always been seen as a beverage of the affluent community (Bisson et al., 2002). Since today, people expect a high quality standard of wine, being a product enjoyable in all sensory aspects, produced in an environmentally sustainable manner, and offering health-promoting properties (Bisson et al., 2002). To guarantee a good wine quality, winemakers need to compete with several challenges that concern horticulture as well as viticulture. A long-standing problem are fungal infections of the grapevine, which may impair wine quality in terms of sensory properties including aroma, flavor, and color.

The European grapevine *Vitis vinifera* L. is mainly used for commercial wine production. It is a perennial plant grown in temperate regions of the world with the capability to withstand environmental stress (Mullins et al., 1992). A number of plant diseases may affect the grapevine, among which fungal pathogens are the most important (Steel et al., 2013). Many parts of the grapevine are susceptible to fungal attacks including the roots, trunk, canes, leaves, and berries (Hocking et al., 2007). Fungal infections of the grapevine may result in large economic losses in wine production, mainly attributed to infections of the berries, which deteriorate grape and wine quality (Steel et al., 2013). The berries are commonly infected by the mildew pathogens *Erysiphe necator* (*Uncinula necator*) and

Plasmopara viticola, as well as *Alternaria* spp., *Aspergillus* spp., *Botrytis cinerea*, *Cladosporium* spp., *Penicillium* spp., and *Rhizopus* spp. (Hocking et al., 2007). Among these, *B. cinerea* is a pathogen frequently found on berries (Steel et al., 2013).

1.1 *Botrytis cinerea* infections

B. cinerea is a ubiquitous, filamentous, and necrotrophic fungus infecting a wide range of plants as a non-host specific organism (Leroux, 2007). It occurs in vineyards all over the world being responsible for the most serious diseases of the grapevine, the gray mold or *Botrytis* bunch rot (Steel et al., 2013). The infection of grapes can also be desired as the noble rot for the production of sweet wines when climatic conditions are favorable, whereas the infection in an early ripening stage with low sugar and high acid concentrations leads to spoilage of the crop (Vortkamp et al., 2013). The gray mold can have several negative consequences on qualities from economical and enological points of view (Ky et al., 2012). It leads to a drastically reduced yield at harvest, which in turn reduces the volume of the wine (Dubos, 1999). The enological quality impact affects the grapes' chemical composition, including major quality compounds such as sugars, organic acids as well as various aroma and phenolic compounds (Ribéreau-Gayon et al., 1980). Organoleptic deteriorations such as mushroom and earthy off-odors have been described (La Guerche et al., 2006). Further problems arise due to the production of a secretome by the fungus, mainly consisting of β-D-glucan, which causes clarification problems (Thakur et al., 2018). The glucan-rich secretome also contains extracellular enzymes as part of the fungal resistance mechanisms against host responses (Gil-ad et al., 2001). During the infection of the grape, *B. cinerea* utilizes several extracellular enzymes to penetrate the grape skin and epidermic layers, and to counteract antifungal agents produced in the grape skin, mainly phenolic compounds (Ky et al., 2012). The stilbene resveratrol is primarily synthesized by the grapevine after fungal infestation and is converted to derivatives like pterostilbene, and δ- and ε-viniferin (Hammerschmidt, 2004). Viniferins are known to exhibit antifungal effects against downy mildew (*P. viticola*) or gray mold (Ehrhardt et al., 2014). Concurrently to the defense of the plant, *B. cinerea* produces laccase to degrade and inactivate antifungal agents. Laccase together with grape tyrosinase represent polyphenoloxidases (PPOs), which are responsible for the quality deterioration of mainly white wine due to oxidative discoloration (Dewey et al.,

2008). The following sections describe important issues of wine quality which are connected to *Botrytis* infections of the grapes, but also to the disease management of fungal infections with copper-based fungicides.

2 Copper in winemaking

The application of copper-containing agents represents one of the oldest approaches to control fungal pathogens in viticulture. Intensive use of copper-based fungicides occurs in the European vineyard since the end of the 19th century to reduce major diseases like downy mildew (Komárek et al., 2010). The Bordeaux mixture was discovered in France in the 1880s. As a simple mixture of copper sulfate and lime, it was the first commercially successful fungicide (Ayres, 2004). Copper is also effective as bactericide, herbicide, and against several other insect pests and diseases (Provenzano et al., 2010). With a greater demand for organic wines, the use of copper- and sulfur-based biopesticides might still increase, since it is the most important pesticide in organic viticulture (Pedneault & Provost, 2016). The perennial application of copper-containing fungicides on the same land together with the low mobility of copper in the soil has led to copper accumulation in vineyard soils and groundwater, causing a risk for human health and the ecosystem (Besnard et al., 2001; Komárek et al., 2010). This problem occurs especially in regions where wet climate enhances the proliferation of fungal diseases (Mirlean et al., 2007). It further raises concerns for fermentation and wine quality, since processing of the grapes with soil copper which is transported over the roots through the transportation system to the grapes or copper attached on the grape surface may result in musts and wines with elevated copper concentrations (Sun et al., 2019). Copper levels in vineyard soils often exceed 200 $mg{\cdot}kg^{-1}$, whereas the warning and critical legislative limits in the EU were set to copper concentrations in agricultural soils of 50 and 140 $mg{\cdot}kg^{-1}$, respectively (Komárek et al., 2010). The determination of a threshold that causes risk is difficult because the mobility and availability of copper are always connected to soil properties such as organic carbon, texture, and pH (Ballabio et al., 2018). An educated use of copper is of great importance for the ecosystem and human health. The European Commission, thus, has limited the annual dose of copper compounds including copper hydroxide, copper oxychloride, copper oxide, Bordeaux mixture, and tribasic copper sulfate to 4 kg copper per ha in the Commission Implementing Regulation (EU) from December 2018.

2.1 Influence of copper on *Saccharomyces cerevisiae*

At cellular level, copper is an essential micronutrient (1–10 µM) required for metabolic processes as a cofactor of enzymes, and is involved in redox reactions and electron transport based on its ability to undergo Cu(I)-to-Cu(II) transitions (Cervantes & Gutierrez-Corona, 1994; Avery et al., 1996). Toxic concentrations, however, influence microbial metabolism and growth due to interactions with nucleic acids, enzyme active sides, and cellular or organellar membranes (Avery et al., 1996). Disruptive effects of copper are especially linked to the induction of oxidative stress within the cells. The copper-mediated Fenton reaction promotes the formation of an elevated cellular level of free hydroxyl radicals ($OH^{\bullet}$) (Shanmuganathan et al., 2004). Major targets of oxidation by $OH^{\bullet}$ and other reactive oxygen species (ROS) are the plasma membrane and proteins (Avery et al., 1996; Shanmuganathan et al., 2004). Free radicals and the redox-active nature of copper lead to a loss of membrane integrity and cell death, usually due to membrane lipid peroxidation. Several studies reported the copper-induced membrane permeabilization in *Saccharomyces cerevisiae*, which is manifested in an efflux of mobile cellular solutes (e.g., K^+) (Avery et al., 1996; Howlett & Avery, 1997). Proteins are affected by ROS due to oxidation of their amino acid side chains inducing hydroxyl or carbonyl derivatives or the cleavage of peptide bonds (Costa et al., 2002; Shanmuganathan et al., 2004). In *S. cerevisiae* high copper exposure predominantly affected catabolic enzymes by specific carbonylation. The majority of proteins susceptible to oxidation were enzymes participating directly in glycolysis or fermentation of the product of glycolysis, pyruvate, but the alcohol dehydrogenase (ADH) (isoform ADHI) has been described to be the most heavily oxidized enzyme (Shanmuganathan et al., 2004). In *S. cerevisiae*, seven genes have been found (ADH1–ADH7) which encode the isoenzymes ADHI to ADHVII (Smidt et al., 2008). Among these, the ADHI is responsible for reducing acetaldehyde to ethanol during fermentation. This was proven with ADHI-defective mutants which showed effects like acetaldehyde accumulation and reduced alcohol production (Ciriacy, 1997). The reversible reduction of acetaldehyde to ethanol generates NAD^+ or NADH depending on the direction (Mauricio et al., 1997). Another impact is the direct substitution of metals at the enzyme active site, which may reduce or inhibit enzyme activities of the yeast. Due to the tetrameric conformation of ADH (150 kDa) that contains structural and catalytic zinc

(Magonet et al., 1992), copper or other transition metals can replace zinc, leading to reduced specific enzyme activity (Cavaletto et al., 2000; Vanni et al., 2002). Copper has led to the highest inhibition efficiency of ADHI in *S. cerevisiae* compared to other transition metals (Cavaletto et al., 2000). All these mechanisms cause serious problems in winemaking when the yeast is exposed to copper stress during fermentation. **Chapter 2** shows the effects of different copper concentrations in the must on two commercially available wine yeasts strains, regarding their fermentation efficiency, vitality, specific ADH activity, and other important fermentation parameters.

2.2 Intracellular copper disposition in *Saccharomyces cerevisiae*

To prevent copper deficiency as well as toxicity, *S. cerevisiae* has developed a complex mechanism of copper trafficking inside the cell to maintain copper homeostasis. It is important to control the transport of this essential micronutrient to its targets, such as enzymes or mitochondria, to prevent spurious interactions. In case of toxic copper amounts, the regulation of the transport and copper sequestration is induced (Vest et al., 2019).

To overcome the plasma membrane as the major barrier of copper distribution in the cell, it contains high- and low-affinity transporters to import copper (Culotta et al., 1999). Copper is transported in its cuprous form, thus, an extracellular metalloreductase is involved in an efficient copper transport via the high-affinity copper transporter 1 (Ctr1) (Yamaguchi-Iwai et al., 1997). Once inside the cell, copper chaperones in the cytoplasm prevent inappropriate interactions by copper-binding and further induce the copper distribution to target enzymes (Vest et al., 2019). This leads to the estimation that the concentration of cytoplasmic free copper is equal to zero. Sequestration of copper is achieved by transport to organelles like the trans-Golgi network, vacuole, or mitochondria (Vest et al., 2019). Besides the chaperone-mediated binding, copper can also be complexed to metallothioneins to prevent redox cycling or uncontrolled binding reactions. In *S. cerevisiae*, two metallothioneins, the Cup1 (cuprum 1) and Crs5 (copper resistant suppressor 5) have been described (Winge et al., 1985; Culotta et al., 1994), but Cup1 is responsible for the majority of copper resistance in yeasts due to its high expression under copper exposure (Jensen et al., 1996). As a consequence of toxic intracellular copper levels in the yeast cell, the transcription of metallothionein genes is, thus, induced and the degradation of the Ctr1 has been described to limit the toxic copper influx (Ooi et al., 1996; Vest et al., 2019). A further important role in heavy metal detoxification can be ascribed to the redox-active tripeptide glutathione (GSH), which is described in section 4.

2.3 Influence of copper on wine quality

Copper may deteriorate wine quality due to a stress-induced stuck and sluggish fermentation. This, however, strongly depends on the copper sensitivity of the yeast strain used (Ferreira et al., 2006). The delayed fermentation reduces alcohol production and increases residual sugar levels in the wine (Bisson, 1999). Furthermore, copper exposition of the yeast during fermentation might increase the volatile acidity of the wine, mainly due to an undesired acetic acid production (Ferreira et al., 2006). Since copper can inhibit important enzymes of *S. cerevisiae* including the ADH, the role of acetaldehyde as an important volatile aroma component in wine needs to be considered. At low levels, acetaldehyde gives a pleasant fruity flavor, but high amounts result in a pungent irritating flavor (Miyake & Shibamoto, 1993; Lachenmeier & Sohnius, 2008). The high affinity to the preservative sulfur dioxide (SO_2) reduces the aroma effect of acetaldehyde and the functionality of SO_2 (Jackowetz et al., 2011). A further influence on wine organoleptic properties arises due to the high reactivity of acetaldehyde with phenolic compounds. It mediates a fast polymerization of anthocyanins and tannins or flavanols, which increases color stability and intensity, but further polymerization with flavanols leads to precipitation, color instability, and color decrease (Es-Safi et al., 2002; Li et al., 2008).

Copper itself is well-known to cause oxidative spoilage of the wine since it catalyzes the first step of non-enzymatic browning in white wine. This follows the same pathways similar to the enzymatic browning described in section 3. Copper oxidizes phenols to reactive *ortho*-quinones, which further polymerize and enhance the browning degree of white wine (Li et al., 2008). Other influences of copper described are the pinking of wine as well as haze formation (Provenzano et al., 2010). This is mainly a problem during wine aging together with a loss of typical wine aromas. An example is the decrease of the strong aroma compound 3-mercaptohexanol, characteristic for a passionfruit odor, during copper exposure of the wine (Ugliano et al., 2011). Copper reactivity with thiols is a known reaction in enology. It is also utilized to remove aromatic defects induced by thiol compounds by adding copper sulfate (Darriet et al., 2001). This is, however, an additional source of elevated copper concentrations in the wine, causing the risk of unwanted side reactions.

3 Influence of polyphenoloxidases on wine color

Wine color is an important quality attribute for the sensory properties of wine. It is usually used as the first clue for wine evaluation giving evidence, for example, of the age or degree of oxidation (Kennedy, 2010). Wine color depends on the chemical composition of the wine, primarily on phenols. Hydroxycinnamates like caffeic, ferulic, and *p*-coumaric acids, as well as their esters with tartaric acid such as caftaric acid, are the principal phenols in white must and wine, with small amounts of flavanols (Waterhouse, 2002). These phenols provide the characteristic yellow color of most white wines (Kennedy 2010). Anthocyanins are responsible for the color of red wines. In red grapes, six anthocyanidins have been identified, with malvidin as the most abundant in *V. vinifera* grapes (Garrido & Borges, 2013). Among the flavonoids, the predominant flavanol in wine is catechin. It is derived from the solid parts of grapes (skin and seeds) (Garrido & Borges, 2013). Seed extraction and extended maceration techniques obtain high levels of catechin (Waterhouse, 2002). The phenolic composition of must and wine strongly depends on winemaking techniques and vinification methods, but also on the grape variety, maturity, and environmental factors, e.g., climate and soil in the vineyards (Fang et al., 2008; Garrido & Borges, 2013). Despite the desired coloring of wine due to its phenolic composition, phenols also contribute to negative color changes throughout the whole winemaking process. The susceptibility of phenols to enzymatic oxidation especially in white must and wine is a well-known problem in winemaking.

3.1 Tyrosinase and laccase

PPOs are copper-containing enzymes responsible for the oxidation of phenols during the processing of grapes in the presence of oxygen. Tyrosinase (*o*-diphenol oxidoreductase, E.C. 1.10.3.1) is a PPO naturally produced by the grape berry, whereas the laccase (*p*-diphenol oxidoreductase, E.C. 1.10.3.2) is a PPO produced and secreted by *B. cinerea* after the infestation of the grape (Dewey et al., 2008). In intact cells of the grape berry, tyrosinase located in the cytoplasm is not in contact with phenolic compounds, which are present in the vacuole, due to different cellular compartments (Wang, 1990). During grape crushing, membrane systems are disrupted and tyrosinase is released into the must, where it catalyzes the hydroxylation of monophenols to *ortho*-diphenols with consecutive oxidation of the catechol to *ortho*-quinones (Claus et al., 2014; Oliveira et al., 2011). These highly reactive species can undergo condensation with other phenolic compounds to form polymerized brown products (Singleton, 1987; Li et al., 2008). In contrast to tyrosinase, laccase causes a larger problem in winemaking regarding quality deterioration due to oxidative browning in white wine and color loss of red wine (Dewey et al., 2008). It is more active than tyrosinase, tolerant against SO_2 and heat treatment usually used in winemaking, stable against alcohol and low pH present in must and wine (Dewey et al., 2008; Ugliano, 2009). Furthermore, the laccase oxidizes a wider range of substrates including mono-, di-, and polyphenols, amino phenols, methoxy phenols, aromatic amines, and ascorbate in a reaction coupled to a four-electron reduction of oxygen to water (Madhavi & Lele, 2009). The first product of oxidation, an unstable free radical, may undergo a second enzyme-catalyzed reaction to form a *para*-quinone (Thurston, 1994; Oliveira et al., 2011). The free radicals as well as the quinones, may polymerize to form brown products (Thurston, 1994). This makes *Botrytis*-infected grapes commonly undesired in winemaking. However, there is strong evidence that different *B. cinerea* isolates do not secrete the same levels of laccases (Cotoras & Silva, 2005) and the degree of infection does not always correspond to laccase activities in the must (Macheix et al., 1991; Perino et al., 1994). **Chapter 3** demonstrates specific laccase activities and the varying catalytic activity of laccase-containing secretomes of different *B. cinerea* strains toward typical wine phenols.

Extracellular *B. cinerea* laccases show a large heterogeneity in their molecular properties. They may occur as different isoenzymes emerging from strain-specific variations, different culture conditions, and purification methods (Claus et al., 2014). Further variations were observed in their inducibility, substrate specificity, molecular weight, isoelectric point, and sugar content as well as temperature and pH optimum (Dubernet et al., 1977; Gigi et al., 1981; Marbach et al., 1984; Zouari et al., 1987; Bollag et al., 1988).

Inducibility of extracellular laccase production may be affected by the addition of grape juice or phenolic compounds, such as gallic and caffeic acids to liquid media (Gigi et al., 1980). Laccase activity *in vitro* is mostly quantified by photometric tests with the use of phenolic substrates by the formation of colored products (Johannes & Majcherczyk, 2000). Common tests utilize the phenolic substrates guaiacol, 2,6-dimethoxyphenol (Prillinger & Esser, 1977), and syringaldazine (Harkin & Obst, 1973), or the non-phenolic substrate 2,2'-azino-bis-(3-ethylbenzothiazoline-6-sulphonic acid) (ABTS) (Childs & Bardsley, 1975). The phenolic substrates form quinones, whereas ABTS results in a colored radical cation (Johannes & Majcherczyk, 2000).

3.2 Enological practices to reduce laccase activity

Some winemaking practices are available to mitigate laccase activity in must and wine. The first step is to minimize the use of infected grapes by selective hand harvesting and manual or automated grape sorting tables in the winery (Steel et al., 2013). Cool conditions at harvest (Wilker, 2010a) followed by gentle and fast transport and immediate processing of the grapes also reduce the effects of laccase (Steel et al., 2013). Common SO_2 levels used in winemaking do not inhibit laccase activity (Dewey et al., 2008), however, high concentrations of SO_2 (100–200 mg·L^{-1}) added to infected fruits or musts (Steel et al., 2013) with concurrent cooling to 10–12 °C might impede oxidation due to the reduction of laccase activity (Wilker, 2010b). Depending on the substrate, *B. cinerea* laccase activity decreases at temperatures above 60 °C (Marbach et al., 1984). Inactivation of laccase in the must, thus, requires pasteurization temperatures exceeding 60 °C, while 80 °C with a holding time of 5 s has been recommended (Steel et al., 2013). Since the laccase requires oxygen to catalyze substrate transformation, a limited exposure of infected grapes to normal atmospheric conditions can control oxidative browning reactions. This can be achieved by the use of inert gas covers during procedures such as pressing, transfers, and for in-tank ullage space (Wilker, 2010a). Further practices such as whole bunch pressing reduce oxygen exposure during white wine production, in contrast to maceration steps of crushing and destemming (Steel et al., 2013). Total oxygen exclusion, however, is not appropriate since yeast starter cultures require small amounts of oxygen during fermentation (Wilker, 2010b). Copper chelating-agents such as EDTA, citric acid, and oxalic acid, as well as the heavy metals copper and cadmium, have been described to serve as laccase inhibitors *in vitro* (Lorenzo et al., 2005), however, their use in winemaking is inappropriate. To avoid the necessity of all these laccase-inhibiting procedures during winemaking, a good disease management is required to inhibit the fungal attack of the grapevine and, thus, reduce the occurrence of laccase in must and wine.

4 Glutathione in winemaking

The lack of suitable alternatives to copper-based fungicides in organic viticulture requires a tool to reduce copper stress for the yeast as well as oxidation in wine. GSH provides a potent natural antioxidative additive in winemaking and simultaneously offers a defense mechanism of the yeast against oxidative stress. Its natural occurrence in the grapes and yeasts presents an ideal requirement for its increased use as an additive to must and wine.

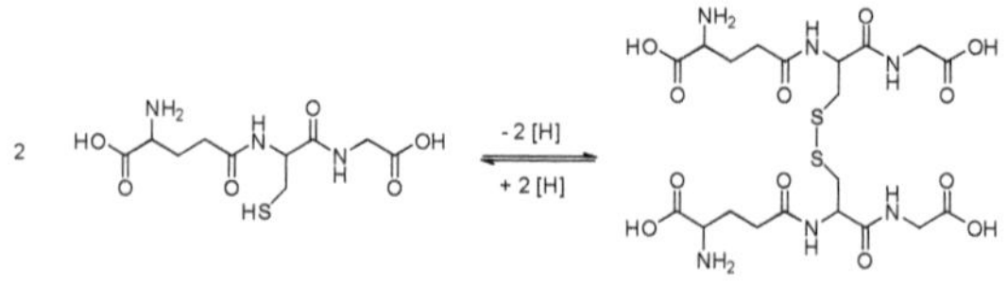

$$2\ \text{GSH} \underset{+2[H]}{\overset{-2[H]}{\rightleftharpoons}} \text{GSSG}$$

Figure 1.1: Reduced glutathione (GSH, left) and the oxidized glutathione disulfide (GSSG, right).

4.1 The role of glutathione at cellular level

GSH is the major non-protein thiol present in high amounts up to 10 mM in most living cells from prokaryotes to eukaryotes (Penninckx, 2000; Penninckx, 2002). The pseudo-tripeptide of L-glutamate, L-cysteine, and glycine is synthesized intracellularly in two ATP-dependent steps by γ-glutamylcysteine synthetase (GSH1) and GSH synthetase (GSH2) (Grant et al., 1996). GSH is generally present with >90% in the reduced form in the cell while the rest occurs in its oxidized form as glutathione disulfide (GSSG) (see **Figure 1.1**) and mixed disulfides, GS-S-Cys and GS-S-CoA (Kritzinger et al., 2013). The NADPH-dependent glutathione reductase can reduce the GSSG back to GSH being again available for the cell (Grant & Dawes, 1996). GSH acts as a strong cellular redox buffer (Penninckx, 2002) and has a strong nucleophilic property related to its thiol moiety of the cysteine residue that protects cells by scavenging free radicals (Grant & Dawes, 1996) and complexation of heavy metals (Połeć-Pawlak et al., 2007). Another detoxification mechanism of the cell is provided by enzymatic defense, where GSH can act as a cofactor (see **Figure 1.2**). This occurs by the conjugation of xenobiotics and less reactive radicals, catalyzed by glutathione S-transferases or the reduction of peroxides by glutathione peroxidase (Hedley & Chow, 1994; Vuilleumier, 1997). If cells are exposed to sulfur or nitrogen starvation, GSH can be used as an endogenous sulfur source or amino acid source to alternatively satisfy growth requirements (Penninckx, 2002). In this case, the γ-glutamyltranspeptidase plays an important role as GSH-degrading enzyme (Mehdi & Penninckx, 1997). The numerous properties of GSH mentioned above, with a pivotal function in bioreduction, protection against oxidative stress, detoxification, enzyme activity as well as sulfur and nitrogen metabolism (Penninckx, 2002) are of great concern for winemaking. As it naturally occurs in grapes and yeasts, it is consequently present in must and wine.

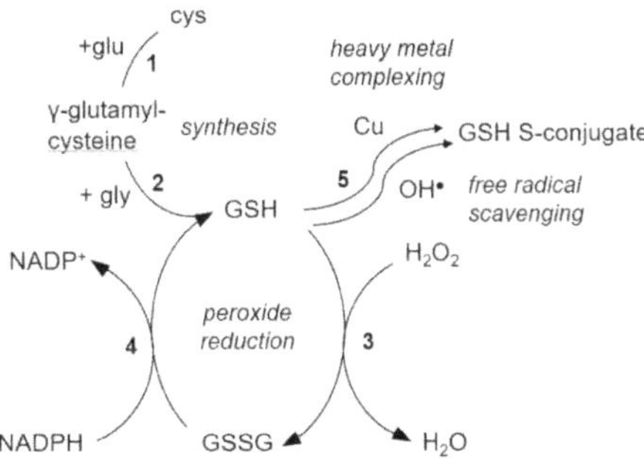

Figure 1.2: Scheme of the protective role of GSH in cells against heavy metals and oxidative stress. **1** γ-glutamylcysteine synthetase, **2** GSH synthetase, **3** GSH peroxidase, **4** GSH reductase **5** GSH S-transferase (Hedley & Chow (1994), modified).

4.2 Glutathione in the grape

The GSH content in grapes is dependent on the grape variety as well as the vintage, location, and technological aspects. This has been reported in a study of 28 *V. vinifera* grape varieties, with a GSH content in the berries ranging from 56 to 372 $\mu mol \cdot kg^{-1}$ (17–114 $mg \cdot kg^{-1}$) (Cheynier et al., 1989). The most important roles of GSH in plants are the redox control, detoxification, and mobilization in case of sulfur and nitrogen starvation (Noctor & Foyer, 1998). Synthesis of GSH occurs in the cytosol and chloroplasts of plant cells. Whether it is synthesized in the berry itself is uncertain, and it is assumed that the source may be the leaves of the grapevine (Kritzinger et al., 2012). It has been observed that an increase in GSH in grape berries of red, green, seeded, and seedless varieties occurs at the onset of ripening (Adams & Liyange, 1993). During this time, >90% of the total glutathione content in the berries was in the reduced form (Okuda and Yokotsuka 1999). There is also evidence that the GSH content in the grapes is associated with the vine nitrogen status, which is estimated as the yeast assimilable nitrogen (YAN) content of the grape must. GSH and YAN levels in the must were much higher after soil nitrogen fertilization, in contrast to must processed from nitrogen-deficient vines (Choné et al., 2006).

4.3 Glutathione in must and wine

GSH levels range from non-detectable to 100 $mg \cdot L^{-1}$ in must (Cheynier et al., 1989; Park et al., 2000b) and from non-detectable to 70 $mg \cdot L^{-1}$ in wine (Fracassetti et al., 2011; Kritzinger, 2012). Factors that may influence GSH concentrations in the must are exposure to oxygen, tyrosinase activity, maceration of grape skin, and pressing (Kritzinger et al., 2012). GSH concentrations in the wine are dependent on the yeast strain used (Lavigne et al., 2007; Kritzinger, 2012) and might be associated with the GSH metabolism of *S. cerevisiae* during fermentation, which utilizes and secretes GSH with the help of transporters, described in section 4.4.1. The complex composition of the must including the initial GSH concentration, as well as the fermentation conditions (temperature, pH, osmotic pressure), may influence yeast growth and, thus, the fluctuation of GSH in the must during fermentation (Kritzinger et al., 2012). During wine aging, GSH levels generally decrease but the concentration can be retained by limiting oxidation during the aging process, for example, by aging wine on their lees (Ugliano et al., 2011; Lavigne et al., 2007).

4.3.1 Glutathione as an antioxidative agent

The antioxidant activity of GSH in must and wine to prevent enzymatic and non-enzymatic oxidative spoilage has been well established. The formation of brown pigments by enzymatic oxidation is reduced in the presence of sufficient amounts of GSH in the must due to the trapping of *o*-quinones which are formed after oxidation of hydroxycinnamates and their esters with tartaric acid, e.g., caftaric and coutaric acids by PPO (Singleton et al., 1985) (see **Figure 1.3**). This leads to the formation of 2-*S*-glutathionyl caftaric acid, known as grape reaction product (GRP) (Cheynier et al., 1986; Salgues et al., 1986). The colorless derivative is not a substrate for grape tyrosinase in spite of its *o*-dihydroxyphenol structure and, thus, the browning of must and wine is reduced (Singleton et al., 1985). However, due to the wider substrate spectrum of laccase, it can oxidize GRP into its corresponding *o*-quinone, and this product further reacts with GSH to form 2,5-di-*S*-glutathionyl caftaric acid (GRP2), or, in the absence of GSH, again proceeds to form brown polymers (Salgues et al., 1986). It has been postulated that the hydroxycinnamic acid-to-GSH ratio of the must should indicate its susceptibility to oxidation (Kritzinger et al., 2012). High ratios (3.8–5.9) were characteristic for dark-colored musts and ratios of 1.1–3.6 and 0.9–2.2 for medium and lightly colored musts, respectively (Du Toit et al., 2006).

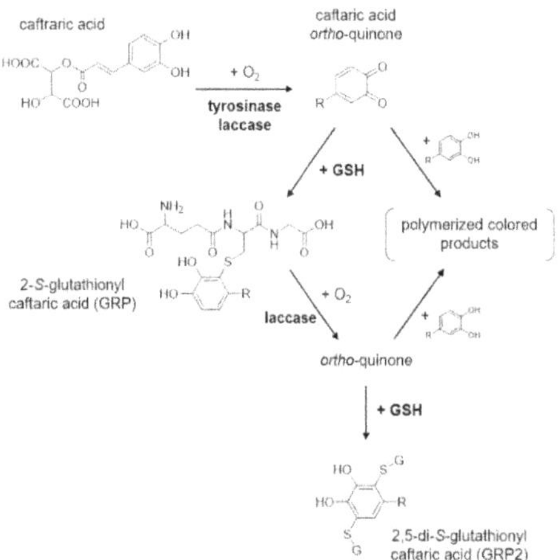

Figure 1.3: Reaction sequence of the oxidation of wine phenols by tyrosinase or laccase, showing the inhibition of polymerization and condensation to colored products by GSH, and the formation of the grape reaction product (GRP) and further oxidation to GRP2 by laccase (Ugliano (2009) and Salgues et al. (1986), modified).

GSH may also play a pivotal role as a chelating agent in copper-rich musts to prevent non-enzymatic oxidation by copper in its free form. This is assumed since GSH can form a mercaptide bond with heavy metals via its thiol group (Cobbett & Goldsbrough, 2002), and others have shown that GSH is previously oxidized to GSSG to form a GSSG-Cu complex (Połeć-Pawlak et al., 2007). The non-enzymatic, oxidative discoloration during wine aging due to the polymerization of oxidized phenols or other wine constituents, such as acetaldehyde or glyoxylic acid (derived from the oxidation of tartaric acid) (Li et al., 2008), might also be reduced by GSH. Caffeic acid and its esters, as well as catechin, epicatechin, and gallic acid have been described to be most susceptible to chemical oxidation in wine (Li et al., 2008). In model white wine the oxidative discoloration induced by the components ascorbic acid, caffeic acid, (+)-catechin, and tartaric acid was reduced by GSH addition (Sonni et al., 2011a). The protective effect of GSH was associated with limitation of *o*-quinones and the delayed formation of carboxymethine-linked (+)-catechin dimers, due to the complexation of carbonyl compounds such as glyoxylic acid (Sonni et al., 2011a; Sonni et al., 2011b). This reduced formation of the intermediate compound, consequently, hinders the formation of yellow xanthylium cations, which might be negatively associated with wine quality (Sonni et al., 2011b).

4.3.2 Glutathione as an aroma preservative

Aroma preservation during aging may be achieved by adding GSH to the wine. The protective effect has been described for various classes of compounds comprising esters and terpenes, contributing to fruity or floral aromas (Papadopoulou & Roussis, 2001; Kritzinger et al., 2012), or volatile thiols such as 3-mercaptohexanol (Ugliano et al., 2011). Volatile thiols and GSH, both having a thiol moiety, might compete to react with *o*-quinones. The trapping of quinones by GSH can preserve thiol-related aromas in wine (Tirelli et al., 2010). The addition of GSH before bottling also inhibited the formation of atypical aging characters linked to sotolon and 2-aminoacetophenone (Dubourdieu & Lavigne-Cruege, 2004; Kritzinger et al., 2012). However, the undesired production of hydrogen disulfide (H_2S), contributing to a reductive, rotten egg-like aroma, may be increased by GSH addition (Kritzinger et al., 2012). In periods of nitrogen and sulfur starvation, GSH is hydrolyzed liberating the breakdown product cysteine as a potent source of H_2S. It has been postulated that GSH is a precursor of 40% of the H_2S liberated by the yeast (Hallinan et al., 1999). GSH addition before bottling also resulted in an elevated H_2S production, especially when high copper concentrations are present in the wine. The authors explain the H_2S accumulation with a low degree of oxidation due to the antioxidant capacity of GSH (Ugliano et al., 2011).

The above described positive and negative effects of GSH addition at different stages of winemaking show the great necessity of further research. The required optimum dose remains unknown as well as the influence of several other parameters during winemaking such as temperature, pH, and SO_2 addition. In 2015, the International Organization of Vine and Wine (OIV) (Oeno 446-2015) allowed the addition of pure reduced GSH to must and wine with a maximum dose of 20 $mg \cdot L^{-1}$. The European Union, however, does not mention GSH in the list of allowed enological treatments (VO EG No. 606/2009). The optimum dose of GSH obviously depends on the natural GSH content of must and wine, which is very variable.

4.3.3 Quantification of glutathione

The lack of strong chromophores or fluorophores in the chemical structure of GSH requires derivatization of the thiol for its quantification in must and wine by liquid chromatography (Camera & Picardo, 2002) The introduction of appropriate labels in the molecule by derivatization improves the detection limit with UV-Vis, allowing fluorescence-based detection (Koller & Eckert, 1997). Because of the high polarity and water solubility of thiols, derivatization allows their separation from complex matrices, and it further enables the detection of low concentrations and stabilizes the highly reactive free sulfhydryl group to prevent errors in analysis (Rao et al., 2010). Requirements for the labeling reagents are high detection sensitivity, specificity and no susceptibility to major matrix interference reactions. GSH offers three sites, the carboxylic, amino, and thiol functional groups, which are all available for derivatization. The preferred site is the thiol moiety to ensure specificity and protection of GSH from auto-oxidation during analysis (Camera and Picardo, 2002). *o*-Phthaldialdehyde (OPA) provides a widely applied reagent for GSH fluorescence detection. OPA is non-fluorescent until it forms an isoindole derivative with thiols in the presence of primary amines (Molnar-Perl, 2001). This heterobifunctional reaction of OPA to derivatize GSH is conducted with a co-reagent such as 2-aminoethanol (Park et al., 2000a). Separation and quantification of a mixture of thiols derivatized with OPA may then be achieved with (U)HPLC and fluorescence detection.

4.4 Glutathione in *Saccharomyces cerevisiae*

The process of fermentation with the use of *S. cerevisiae* presents a crucial subject for technologists because the yeast is frequently exposed to stress occurring before, during, and after fermentation. The challenge is to maintain metabolic vitality and viability of the yeast to prevent an impairment of the fermentation and, thus, ensure a good wine quality. *S. cerevisiae* exhibits different response mechanisms toward commonly occurring stress: pH and temperature shocks, oxidative and osmotic stress, and toxic fermentation products, mainly ethanol. Heat shock proteins and compatible solutes such as glycerol and trehalose are produced against thermal or osmotic shock, or other stresses like ethanol (Penninckx, 2000). Ethanol resistance also increases by a change in membrane composition with an increased percentage of unsaturated fatty acids and a change in sterol content (Arneborg et al., 1995). In case of oxidative and nutritional stresses, glutathione-dependent systems are of great importance (Penninckx, 2000). Depending on the growth conditions, GSH comprises 0.5 to 1% of the dry weight in *S. cerevisiae* (Penninckx, 2002). It is of great interest to generate yeast strains with an enhanced GSH production capability. They can be applied either as active yeast strains to increase the GSH content in the wine or as inactivated dry yeast (IDY) preparations added to the wine. Approaches to obtain active yeast with an enhanced GSH production were genetically engineered strains, which overexpress enzymes of GSH biosynthesis, or enzymes involved in sulfur assimilation for an increased cysteine biosynthesis (Grant et al., 1997; Hara et al., 2012). Others were mutation strategies performed by chemical and physical treatments (Lai et al., 2008; Chen et al., 2012), or evolution-based strategies (Mezzetti et al., 2014). IDY preparations are already commonly used products within the enological industry. They are obtained from thermal inactivation of the yeast grown in highly concentrated sugar medium under aerobic conditions and are available in several forms as inactive yeast, yeast autolysates, yeast extracts, and yeast hulls or walls (Pozo-Bayón et al., 2009). Traditionally, IDY were applied to improve fermentation by protecting the active dry yeast during rehydration and enhancing alcoholic fermentation as a nutrient source or an adsorber of toxic compounds (Pozo-Bayón et al., 2009). Their use to enhance wine sensory properties due to a high GSH content has recently received attention. Commercially available products claim the preservation of wine aroma, extension of shelf life after bottling, delay of the development of oxidized notes and the yellowing of white wine (Kritzinger et al., 2012). GSH-enriched dry yeast (GSH-IDY) preparations increase the wine GSH content by secretion of GSH into the wine or an increased GSH production by the yeast due to the assimilation of GSH precursors (Kritzinger et al., 2012). In 2017, the OIV (Oeno 533-2017) has set the treatment with IDY to guaranteed GSH levels of 20 $mg \cdot L^{-1}$.

4.4.1 Glutathione assimilation and secretion

Additionally to the endogenous GSH biosynthesis, *S. cerevisiae* may assimilate GSH from the extracellular medium via specific transporters. This makes an increased GSH content in wine by the application of GSH-IDY or pure GSH of great interest in the protection of cells from oxidative stress and the support of growth conditions. For *S. cerevisiae*, a high-affinity plasma membrane GSH transporter Hgt1p has been characterized (Bourbouloux et al., 2000) and two kinetically distinct GSH transport systems: the GSH-P1 high-affinity ATP-driven and regulated transporter, active in case of sulfur starvation, and the unregulated GSH-P2 transporter with a lower GSH affinity (Miyake et al., 1998). Another transporter (Gex1) has been described for *S. cerevisiae* also responsible for the secretion of GSH to the extracellular medium to maintain GSH homeostasis and for cadmium detoxification (Dhaoui et al., 2011), a defense mechanism that is well documented. GSH spontaneously reacts with cadmium, other heavy metals or xenobiotics, or via the GSH S-transferase to form stable GSH S-conjugates (Połeć-Pawlak et al., 2007; Vuilleumier, 1997). The complex is then transported by GSH S-conjugate export pumps into the vacuolar lumen or in the extracellular medium (Dhaoui et al., 2011). For *S. cerevisiae*, the vacuolar membrane protein Ycf1, the plasma membrane protein Yor1 as well as the Gex1 protein located at both sites have been identified to play a special role in cadmium detoxification (Li et al., 1996; Nagy et al., 2006; Dhaoui et al., 2011). The uptake of GSH and its necessity in *S. cerevisiae* for metabolic processes and cell protection against oxidative stress has been proven by the generation of mutants with a gene disruption in *GSH1*, which were not able to produce the first enzyme γ-glutamylcysteine synthetase in GSH biosynthesis (Grant et al., 1996). In the absence of GSH, yeasts were not able to survive and grow on minimal medium due to the requirement for GSH as a reductant and showed a hypersensitivity against ROS (e.g., H_2O_2). Growth of the mutants could be restored by exogenous GSH addition (Grant et al., 1996).

5 Alternative antioxidative agents and techniques in winemaking

The preservation of must and wine against oxidative degradation is crucial for wine quality especially during an extended storage time. Phenolic compounds in wine already provide a high antioxidant potential because of their ability of free radical scavenging. White wines generally have a lower antioxidative potential than red wines since the maceration step during the making of red wines results in higher levels of phenolic compounds (Abramovič et al., 2015). Total polyphenol contents of red wines range from 300–5000 $mg \cdot L^{-1}$, while concentrations in white wine are in the range of 60–200 $mg \cdot L^{-1}$ (Claus et al., 2014). Consequently, white wines are more prone to oxidation and require the use of antioxidative additives. Besides GSH, other preservatives exist to protect must and wine from oxidation, with SO_2 as the most common, always considering the exposure to oxygen, the activator for enzymatic and chemical oxidation.

5.1 Sulfur dioxide

The preservative SO_2 displays antioxidative and antimicrobial properties. It is naturally present in wine due to its production by the yeast during fermentation (Ribéreau-Gayon et al., 2006). Its addition to must and wine can be protective throughout various stages of winemaking, from pressing to bottling, especially in white wine (Oliveira et al., 2011). During grape crushing, SO_2 prevents browning by inactivation of tyrosinase, or to a lesser extent laccase and may inhibit the growth of spoilage yeasts or bacteria (Du Toit et al., 2006; Santos et al., 2012). Later addition before bottling prevents oxidative spoilage due to phenolic polymerization during wine aging (Santos et al., 2012). The antioxidative property of SO_2 is based on the reaction with H_2O_2, the scavenging of *o*-quinones and their reduction back to the initial phenolic compounds (Santos et al., 2012; Barril et al., 2012). However, the addition of SO_2 raises health concerns because it may cause pseudo-allergic reactions in sulfite-sensitive individuals (Li et al., 2008). This led to a progressive reduction of the allowed SO_2 concentrations advised in wines by the OIV

(Santos et al., 2012). Excessive SO_2 concentrations in must and wine also adversely affect wine quality, leading to unpleasant flavors and aromas, and may cause a cloudy appearance during storage (Li et al., 2008). Addition of precise quantities of SO_2 is therefore necessary.

5.2 Ascorbic acid

Ascorbic acid has also been widely used as an antioxidative agent in winemaking, alternatively or in combination with SO_2, since ascorbic acid does not display antimicrobial properties (Ribéreau-Gayon et al., 2006). It directly reacts with molecular oxygen with a preferential reactivity over phenolic compounds and reduces the formed *o*-quinones (Barril et al., 2012; Abramovič et al., 2015). But the oxidation of ascorbic acid generates H_2O_2, which leads to spoilage reactions, and the formation of dehydroascorbic acid that degrades into a wide range of products (Barril et al., 2012). Many studies have shown that the application of ascorbic acid in combination with SO_2 accelerated the consumption of SO_2 and the production of yellow pigments (Li et al., 2008). Consequently, ascorbic acid is less widely used in winemaking and is even not approved in France (Ribéreau-Gayon et al., 2006).

5.3 Aging on lees

A similar approach like the use of GSH and GSH-IDY is the wine aging on yeast lees. This method utilizes the lees' ability of oxygen consumption. It has been reported that it contributes to the improvement of organoleptic properties, such as stabilizing color and diminishing aroma degradation (Pérez-Serradilla & Castro, 2008). The release of several compounds including peptides, amino acids, sterols, and polysaccharides (mannoproteins) by the yeast during autolysis might be responsible. Because it is a slow and very complex process with the risk of microbiological and organoleptic defects, the application of IDY preparations has gained much more attention (Pozo-Bay6n et al., 2009).

6 Aims of the thesis

Fungal infections of the grapevine are still causing a vast disease problem, confronting the wine industry all over the world. In times of climate change, with unusual and unpredictable weather conditions and an ongoing emergence of resistant fungal strains, the disease management is strongly impeded. The traditional use of copper-based fungicides is still a commonly applied method to control major diseases like downy and powdery mildew, or the gray mold. This alternative to synthetic pesticides is mostly applied in organic viticulture, which has experienced an increased demand in the past years. However, the application of this toxic heavy metal in the vineyard introduces health concerns and might deteriorate wine quality. Copper residues on the grape skin and the accumulated soil copper which is transported over the roots into the grapes result in copper-rich musts. The effects of copper exposition to the wine yeast *Saccharomyces cerevisiae* are well known. If the essential micronutrient reaches concentrations that the yeasts' detoxification mechanism is unable to cope with, reductions in cell vitality and important enzyme activities appear. This knowledge led to the aim of this work, to analyze the effects of copper on the yeast fermentation efficiency and alcohol dehydrogenase activity. This enzyme is crucial for alcoholic fermentation since it catalyzes the last rate-limiting step from acetaldehyde to ethanol. Previous studies showed that, depending on the copper resistance, strains exhibited different thresholds leading to copper stress. In this thesis, two yeast strains were used and different copper concentrations (2.5 to 25 $mg \cdot L^{-1}$) were added to the must before inoculation. The evaluation of copper-induced stress was accompanied by the addition of 20 $mg \cdot L^{-1}$ reduced glutathione (GSH) to counteract copper stress. GSH is an important cellular defense mechanism of the yeast against oxidative and heavy metal stress. It is supposed to be an optimal natural additive to copper-rich must due to its protective effect either in the must as well as inside the yeast cell. However, its application is still discussed controversially and factors influencing its functionality in must and wine are not fully understood. The use of GSH in combination with copper should give new information on the application of GSH in winemaking.

The reduction of fungal pathogens in the vineyard is of great concern for wine quality. Wines prepared from *Botrytis cinerea*-infected grapes have an increased risk to exhibit organoleptic defects. This is mainly a result of the enzyme laccase, a polyphenoloxidase, which enhances enzymatic browning reactions, especially in white wines. It has been shown that must prepared from grapes with a high degree of infection do not necessarily exhibit high laccase activities. Therefore, this work aimed to gain a deeper knowledge of the laccase activity in the secretomes of 10 different *B. cinerea* isolates from red and white grape varieties. To obtain the laccase-containing secretomes, strains were cultivated in the presence of gallic acid, an effective inducer of laccase production. Several phenols contribute to the color deterioration of wine due to their susceptibility to oxidation by laccase. Thus, the substrates applied included various classes of wine phenolic compounds. Identification of oxidation products, as well as their product formation rates, relative product concentrations, and substrate half-life, should give detailed information on the catalytic activity of strain-specific laccases. This is an important approach to better understand and control color deterioration by laccases.

The following specific aims were included in two individual studies (**Chapters 2** and **3**).

- Microvinifications with two yeast strains in Riesling must enriched with different copper concentrations with and without the addition of GSH (**Chapter 2**)
- Determination of yeast vitality and fermentation efficiency (total sugar degradation) as well as acetaldehyde accumulation, GSH contents, and copper concentrations in the must (**Chapter 2**)
- Ultrasound disruption of yeast cells to obtain cell-free extracts for the determination of specific alcohol dehydrogenase activity and intracellular copper concentrations (**Chapter 2**)
- Cultivation of 10 different *B. cinerea* isolates from red and white grape varieties in the presence of the laccase-inducing agent gallic acid to obtain the laccase-containing secretomes (**Chapter 3**)
- Determination of specific laccase activity using the syringaldazine assay (**Chapter 3**)
- Comparison of different secretomes in their catalytic activity toward wine phenols by UHPLC-DAD quantification of oxidation products over time and determination of substrate half-life (**Chapter 3**)
- Identification of oxidation products by UHPLC-ESI-MS^n (**Chapter 3**)

References

Abramovič, H.; Košmerl, T.; Ulrih, N. P.; Cigić, B. Contribution of SO_2 to antioxidant potential of white wine. *Food Chemistry*, **2015**, 174, 147–153.

Adams D. O.; Liyange, C. Glutathione increases in grape berries at the onset of ripening. *American Journal of Enology and Viticulture*, **1993**, 44, 333–338.

Arneborg, N.; Høy, C.-E.; Jørgensen, O. B. The effect of ethanol and specific growth rate on the lipid content and composition of *Saccharomyces cerevisiae* grown anaerobically in a chemostat. *Yeast*, **1995**, 11, 953–959.

Avery, S. V.; Howlett, N. G.; Radice, S. Copper toxicity towards *Saccharomyces cerevisiae*: dependence on plasma membrane fatty acid composition. *Applied and Environmental Microbiology*, **1996**, 62, 3960–3966.

Ayres, P. G. Alexis Millardet: France's forgotten mycologist. *Mycologist*, **2004**, 18, 23–26.

Ballabio, C.; Panagos, P.; Lugato, E.; Huang, J.-H.; Orgiazzi, A.; Jones, A.; Fernández-Ugalde, O.; Borrelli, P.; Montanarella, L. Copper distribution in European topsoils: An assessment based on LUCAS soil survey. *Science of the Total Environment*, **2018**, 636, 282–298.

Barril, C.; Clark, A. C.; Scollary, G. R. Chemistry of ascorbic acid and sulfur dioxide as an antioxidant system relevant to white wine. *Analytica Chimica Acta*, **2012**, 732, 186–193.

Besnard, E.; Chenu, C.; Robert, M. Influence of organic amendments on copper distribution among particle-size and density fractions in Champagne vineyard soils. *Environmental Pollution*, **2001**, 112, 329–337.

Bisson, L. F. Stuck and sluggish fermentations. *American Journal of Enology and Viticulture*, **1999**, 50, 107–119.

Bisson, L. F.; Waterhouse, A. L.; Ebeler, S. E.; Walker, M. A.; Lapsley, J. T. The present and future of the international wine industry. *Nature*, **2002**, 418, 696–699.

Bollag, J. M.; Shuttleworth, K. L.; Anderson, D. H. Laccase-mediated detoxification of phenolic compounds. *Applied and Environmental Microbiology*, **1988**, 54, 3086–3091.

Bourbouloux, A.; Shahi, P.; Chakladar, A.; Delrot, S.; Bachhawat, A. K. Hgt1p, a high affinity glutathione transporter from the yeast *Saccharomyces cerevisiae*. *Journal of Biological Chemistry*, **2000**, 275, 13259–13265.

Camera, E.; Picardo, M. Analytical methods to investigate glutathione and related compounds in biological and pathological processes. *Journal of Chromatography. B*, **2002**, 781, 181–206.

Cavaletto, M.; Pessione, E.; Vanni, A.; Giunta, C. Improved resistance to transition metals of a cobalt-substituted alcohol dehydrogenase 1 from *Saccharomyces cerevisiae*. *Journal of Biotechnology*, **2000**, 84, 87–91.

Cervantes, C.; Gutierrez-Corona, F. Copper resistance mechanisms in bacteria and fungi. *FEMS Microbiology Reviews*, **1994**, 14, 121–137.

Chen, Y.; Yang, X.; Zhang, S.; Wang, X.; Guo, C.; Guo, X.; Xiao, D. Development of *Saccharomyces cerevisiae* producing higher levels of sulfur dioxide and glutathione to improve beer flavor stability. *Applied Biochemistry and Biotechnology*, **2012**, 166, 402–413.

Cheynier, V.; Souquet, J. M.; Moutounet, M. Glutathione content and glutathione to hydroxycinnamic acid ratio in *Vitis vinifera* grapes and musts. *American Journal of Enology and Viticulture*, **1989**, 40, 320–324.

Cheynier, V. F.; Trousdale, E. K.; Singleton, V. L.; Salgues, M. J.; Wylde, R. Characterization of 2-*S*-glutathionyl caftaric acid and its hydrolysis in relation to grape wines. *Journal of Agricultural and Food Chemistry*, **1986**, 34, 217–221.

Childs, R. E.; Bardsley, W. G. The steady-state kinetics of peroxidase with 2, 2'-azino-di-(3-ethyl-benzthiazoline-6-sulphonic acid) as chromogen. *Biochemical Journal*, **1975**, 145, 93.

Choné, X.; Lavigne-Cruège, V.; Tominaga, T.; van Leeuwen, C.; Castagnède, C.; Saucier, C.; Dubourdieu, D. Effect of vine nitrogen status on grape aromatic potential: flavor precursors (S-cysteine conjugates), glutathione and phenolic content in Vitis vinifera L. Cv Sauvignon blanc grape juice. *Oeno One*, **2006**, 40, 1–6.

Ciriacy, M. Alcohol dehydrogenases. In *Yeast sugar metabolism. Biochemistry, genetics, biotechnology and applications*; Zimmerman, F. K.; Entian, K.-D., Eds.; Technomic Publication: Pennsylvania, USA, 1997, 213–223.

Claus, H.; Sabel, A.; König, H. Wine phenols and laccase: An Ambivalent Relationship. In *Wine phenolic composition, classification and health benefits*; El Rayess, Y., Ed.; Nova Publishers: New York, USA, 2014, 155–185.

Cobbett, C.; Goldsbrough, P. Phytochelatins and metallothioneins: roles in heavy metal detoxification and homeostasis. *Annual Review of Plant Biology*, **2002**, 53, 159–182.

Commission Implementing Regulation (EU) 2018/1981 of 13 December 2018 renewing the approval of the active substances copper compounds, as candidates for substitution, in accordance with Regulation (EC) No 1107/2009 of the European Parliament and of the Council concerning the placing of plant protection products on the market, and amending the Annex to Commission Implementing Regulation (EU) No 540/2011. *Official Journal of the European Union* L 317, 14.12.2018, 16–20.

Costa, V. M. V.; Amorim, M. A.; Quintanilha, A.; Moradas-Ferreira, P. Hydrogen peroxide-induced carbonylation of key metabolic enzymes in *Saccharomyces cerevisiae*: the involvement of the oxidative stress response regulators Yap1 and Skn7. *Free Radical Biology and Medicine*, **2002**, 33, 1507–1515.

Cotoras, M.; Silva, E. Differences in the initial events of infection of Botrytis cinerea strains isolated from tomato and grape. *Mycologia*, **2005**, 97, 485–492.

Culotta, V. C.; Howard, W. R.; Liu, X. F. CRS5 encodes a metallothionein-like protein in *Saccharomyces cerevisiae*. *Journal of Biological Chemistry*, **1994**, 269, 25295–25302.

Culotta, V. C.; Lin, S.-J.; Schmidt, P.; Klomp, L. W. J.; Casareno, R. L. B.; Gitlin, J. Intracellular pathways of copper trafficking in yeast and humans. In *Copper transport and its disorders, Molecular and cellular aspects*; Leone, A.; Mercer, J. F. B., Eds.; Kluwer Academic / Plenum Publisher: New York, USA, 1999, 448, 247–254.

Darriet, P.; Bouchilloux, P.; Poupot, C.; Bugaret, Y.; Clerjeau, M.; Sauris, P.; Medina, B.; Dubourdieu, D. Effects of copper fungicide spraying on volatile thiols of the varietal aroma of Sauvignon blanc, Cabernet Sauvignon and Merlot wines. *Vitis*, **2001**, 40, 93–100.

Dewey, F. M.; Hill, M.; DeScenzo, R. Quantification of *Botrytis* and laccase in winegrapes. *American Journal of Enology and Viticulture*, **2008**, 59, 47–54.

Dhaoui, M.; Auchère, F.; Blaiseau, P.-L.; Lesuisse, E.; Landoulsi, A.; Camadro, J.-M.; Haguenauer Tsapis, R.; Belgareh Touzé, N. Gex1 is a yeast glutathione exchanger that interferes with pH and redox homeostasis. *Molecular Biology of the Cell*, **2011**, 22, 2054–2067.

Du Toit, W. J.; Marais, J.; Pretorius, I. S.; Du Toit, M. Oxygen in must and wine: A review. *South African Journal of Enology and Viticulture*, **2006**, 27, 76–94.

Dubernet, M.; Ribereau-Gayon, P.; Lerner, H. R.; Harel, E.; Mayer, A. M. Purification and properties of laccase from *Botrytis cinerea*. *Phytochemistry*, **1977**, 16, 191–193.

Dubos, B. La porriture grise. In *Maladies cryptogamiques de la vigne: champignons parasites des organes herbacés et du bois de la vigne*; Dubos, B., Ed.; Féret: Bordeaux, France, 1999, 55–71.

Dubourdieu, D.; Lavigne-Cruege, V. The role of glutathione on the aromatic evolution of dry white wine. *Vinidea.net Wine Internet Technical Journal*, **2004**, 2, 1–9.

Ehrhardt, C.; Arapitsas, P.; Stefanini, M.; Flick, G.; Mattivi, F. Analysis of the phenolic composition of fungus-resistant grape varieties cultivated in Italy and Germany using UHPLC-MS/MS. *Journal of Mass Spectrometry*, **2014**, 49, 860–869.

Es-Safi, N.-E.; Cheynier, V.; Moutounet, M. Role of aldehydic derivatives in the condensation of phenolic compounds with emphasis on the sensorial properties of fruit-derived foods. *Journal of Agricultural and Food Chemistry*, **2002**, 50, 5571–5585.

Fang, F.; Li, J.-M.; Zhang, P.; Tang, K.; Wang, W.; Pan, Q.-H.; Huang, W.-d. Effects of grape variety, harvest date, fermentation vessel and wine ageing on flavonoid concentration in red wines. *Food Research International*, **2008**, 41, 53–60.

Ferreira, J.; Du Toit, M.; Du Toit, W. J. The effects of copper and high sugar concentrations on growth, fermentation efficiency and volatile acidity production of different commercial wine yeast strains. *Australian Journal of Grape and Wine Research*, **2006**, 12, 50–56.

Fracassetti, D.; Lawrence, N.; Tredoux, A. G. J.; Tirelli, A.; Nieuwoudt, H. H.; Du Toit, W. J. Quantification of glutathione, catechin and caffeic acid in grape juice and wine by a novel ultra-performance liquid chromatography method. *Food Chemistry*, **2011**, 128, 1136–1142.

Garrido, J.; Borges, F. Wine and grape polyphenols - A chemical perspective. *Food Research International*, **2013**, 54, 1844–1858.

Gigi, O.; Marbach, I.; Mayer, A. M. Induction of laccase formation in *Botrytis*. *Phytochemistry*, **1980**, 19, 2273–2275.

Gigi, O.; Marbach, I.; Mayer, A. M. Properties of gallic acid-induced extracellular laccase of *Botrytis cinerea*. *Phytochemistry*, **1981**, 20, 1211–1213.

Gil-ad, N. L.; Bar-Nun, N.; Mayer, A. M. The possible function of the glucan sheath of *Botrytis cinerea*: effects on the distribution of enzyme activities. *FEMS Microbiology Letters*, **2001**, 199, 109–113.

Grant, C. M.; Dawes, I. W. Synthesis and role of glutathione in protection against oxidative stress in yeast. *Redox Report*, **1996**, 2, 223–229.

Grant, C. M.; MacIver, F. H.; Dawes, I. W. Glutathione is an essential metabolite required for resistance to oxidative stress in the yeast *Saccharomyces cerevisiae*. *Current Genetics*, **1996**, 29, 511–515.

Grant, C. M.; MacIver, F. H.; Dawes, I. W. Glutathione synthetase is dispensable for growth under both normal and oxidative stress conditions in the yeast *Saccharomyces cerevisiae* due to an accumulation of the dipeptide gamma-glutamylcysteine. *Molecular Biology of the Cell*, **1997**, 8, 1699–1707.

Hallinan, C. P.; Saul, D. J.; Jiranek, V. Differential utilisation of sulfur compounds for H_2S liberation by nitrogen-starved wine yeasts. *Australian Journal of Grape and Wine Research*, **1999**, 5, 82–90.

Hammerschmidt, R. The metabolic fate of resveratrol: key to resistance in grape? *Physiological and Molecular Plant Pathology*, **2004**, 6, 269–270.

Hara, K. Y.; Kiriyama, K.; Inagaki, A.; Nakayama, H.; Kondo, A. Improvement of glutathione production by metabolic engineering the sulfate assimilation pathway of *Saccharomyces cerevisiae*. *Applied Microbiology and Biotechnology*, **2012**, 94, 1313–1319.

Harkin, J. M.; Obst, J. R. Syringaldazine, an effective reagent for detecting laccase and peroxidase in fungi. *Experientia*, **1973**, 29, 381–387.

Hedley, D. W.; Chow, S. Evaluation of methods for measuring cellular glutathione content using flow cytometry. *Cytometry*, **1994**, 15, 349–358.

Hocking, A. D.; Su-lin, L. L.; Kazi, B. A.; Emmett, R. W.; Scott, E. S. Fungi and mycotoxins in vineyards and grape products. *International Journal of Food Microbiology*, **2007**, 119, 84–88.

Howlett, N. G.; Avery, S. V. Induction of lipid peroxidation during heavy metal stress in *Saccharomyces cerevisiae* and influence of plasma membrane fatty acid unsaturation. *Applied and Environmental Microbiology*, **1997**, 63, 2971–2976.

Jackowetz, J. N.; Dierschke, S.; Orduña, R. M. de Multifactorial analysis of acetaldehyde kinetics during alcoholic fermentation by *Saccharomyces cerevisiae*. *Food Research International*, **2011**, 44, 310–316.

Jensen, L. T.; Howard, W. R.; Strain, J. J.; Winge, D. R.; Culotta, V. C. Enhanced effectiveness of copper ion buffering by CUP1 metallothionein compared with CRS5 metallothionein in *Saccharomyces cerevisiae*. *Journal of Biological Chemistry*, **1996**, 271, 18514–18519.

Johannes, C.; Majcherczyk, A. Laccase activity tests and laccase inhibitors. *Journal of Biotechnology*, **2000**, 78, 193–199.

Kennedy, J. A. Wine colour. In *Viticulture and wine quality*; Reynolds, A. G., Ed.; Woodhead Publishing Limited: Cambridge, UK, 2010, Vol. 1, 73–104.

Koller, M.; Eckert, H. Derivatization of peptides for their determination by chromatographic methods. *Analytica Chimica Acta*, **1997**, 352, 31–59.

Komárek, M.; Čadková, E.; Chrastný, V.; Bordas, F.; Bollinger, J.-C. Contamination of vineyard soils with fungicides: a review of environmental and toxicological aspects. *Environment International*, **2010**, 36, 138–151.

Kritzinger, E. C. Winemaking practices affecting glutathione concentrations in white wine. M.Sc. Thesis, Stellenbosch University, Matieland, South Africa, 2012.

Kritzinger, E. C.; Bauer, F. F.; Du Toit, W. J. Role of Glutathione in Winemaking: A Review. *Journal of Agricultural and Food Chemistry*, **2012**, 61, 269–277.

Kritzinger, E. C.; Stander, M. A.; Du Toit, W. J. Assessment of glutathione levels in model solution and grape ferments supplemented with glutathione-enriched inactive dry yeast preparations using a novel UPLC-MS/MS method. *Food Additives & Contaminants: Part A*, **2013**, 30, 80–92.

Ky, I.; Lorrain, B.; Jourdes, M.; Pasquier, G.; Fermaud, M.; Gény, L.; Rey, P.; Doneche, B.; Teissedre, P.-L. Assessment of grey mould (*Botrytis cinerea*) impact on phenolic and sensory quality of Bordeaux grapes, musts and wines for two consecutive vintages. *Australian Journal of Grape and Wine Research*, **2012**, 18, 215–226.

La Guerche, S.; Dauphin, B.; Pons, M.; Blancard, D.; Darriet, P. Characterization of some mushroom and earthy off-odors microbially induced by the development of rot on grapes. *Journal of Agricultural and Food Chemistry*, **2006**, 54, 9193–9200.

Lachenmeier, D. W.; Sohnius, E.-M. The role of acetaldehyde outside ethanol metabolism in the carcinogenicity of alcoholic beverages: evidence from a large chemical survey. *Food and Chemical Toxicology*, **2008**, 46, 2903–2911.

Lai, J.-T.; Lee, S.-Y.; Hsieh, C.-C.; Hwang, C.-F.; Liao, C.-C. *Saccharomyces cerevisiae* strains for hyper-producing glutathione and γ-glutamylcysteine and processes of use. *US Patent. 7,371,557, p. B2*, **2008**.

Lavigne, V.; Pons, A.; Dubourdieu, D. Assay of glutathione in must and wines using capillary electrophoresis and laser-induced fluorescence detection: changes in concentration in dry white wines during alcoholic fermentation and aging. *Journal of Chromatography A*, **2007**, 1139, 130–135.

Leroux, P. Chemical control of *Botrytis* and its resistance to chemical fungicides. In *Botrytis: Biology, pathology and control*; Elad, Y.; Williamson, B.; Tudzynski, P.; Delen, N., Eds.; Springer: Dordrecht, The Netherlands, 2007, 195–222.

Li, H.; Guo, A.; Wang, H. Mechanisms of oxidative browning of wine. *Food Chemistry*, **2008**, 108, 1–13.

Li, Z.-S.; Szczypka, M.; Lu, Y.-P.; Thiele, D. J.; Rea, P. A. The yeast cadmium factor protein (YCF1) is a vacuolar glutathione S-conjugate pump. *Journal of Biological Chemistry*, **1996**, 271, 6509–6517.

Lorenzo, M.; Moldes, D.; Couto, S. R.; Sanromán, M. A.A. Inhibition of laccase activity from *Trametes versicolor* by heavy metals and organic compounds. *Chemosphere*, **2005**, 60, 1124–1128.

Macheix, J.-J.; Sapis, J.-C.; Fleuriet, A.; Lee, C. Y. Phenolic compounds and polyphenoloxidases in relation to browning in grapes and wine. *Critical Reviews in Food Science & Nutrition*, **1991**, 30, 441–486.

Madhavi, V.; Lele, S. S. Laccase. Properties and applications. *BioResources*, **2009**, 4, 1694–1717.

Magonet, E.; Hayen, P.; Delforge, D.; Delaive, E.; Remacle, J. Importance of the structural zinc atom for the stability of yeast alcohol dehydrogenase. *Biochemical Journal*, **1992**, 287, 361–365.

Marbach, I.; Harel, E.; Mayer, A. M. Molecular properties of extracellular *Botrytis cinerea* laccase. *Phytochemistry*, **1984**, 23, 2713–2717.

Mauricio, J. C.; Moreno, J. J.; Ortega, J. M. In vitro specific activities of alcohol and aldehyde dehydrogenases from two flor yeasts during controlled wine aging. *Journal of Agricultural and Food Chemistry*, **1997**, 45, 1967–1971.

Mehdi, K.; Penninckx, M. J. An important role for glutathione and γ-glutamyltranspeptidase in the supply of growth requirements during nitrogen starvation of the yeast *Saccharomyces cerevisiae*. *Microbiology*, **1997**, 143, 1885–1889.

Mezzetti, F.; Vero, L. de; Giudici, P. Evolved *Saccharomyces cerevisiae* wine strains with enhanced glutathione production obtained by an evolution-based strategy. *FEMS Yeast Research*, **2014**, 14, 977–987.

Mirlean, N.; Roisenberg, A.; Chies, J. O. Metal contamination of vineyard soils in wet subtropics (southern Brazil). *Environmental Pollution*, **2007**, 149, 10–17.

Miyake, T.; Hazu, T.; Yoshida, S.; Kanayama, M.; Tomochika, K.-i.; Shinoda, S.; Ono, B.-i. Glutathione transport systems of the budding yeast *Saccharomyces cerevisiae*. *Bioscience, Biotechnology, and Biochemistry*, **1998**, 62, 1858–1864.

Miyake, T.; Shibamoto, T. Quantitative analysis of acetaldehyde in foods and beverages. *Journal of Agricultural and Food Chemistry*, **1993**, 41, 1968–1970.

Molnar-Perl, I. Derivatization and chromatographic behavior of the o-phthaldialdehyde amino acid derivatives obtained with various SH-group-containing additives. *Journal of Chromatography A*, **2001**, 913, 283–302.

Mullins, M. G.; Bouquet, A.; Williams, L. E. *Biology of the grapevine*; Cambridge University Press: Cambridge, UK, 1992.

Nagy, Z.; Montigny, C.; Leverrier, P.; Yeh, S.; Goffeau, A.; Garrigos, M.; Falson, P. Role of the yeast ABC transporter Yor1p in cadmium detoxification. *Biochimie*, **2006**, 88, 1665–1671.

Noctor, G.; Foyer, C. H. Simultaneous measurement of foliar glutathione, γ-glutamylcysteine, and amino acids by high-performance liquid chromatography: comparison with two other assay methods for glutathione. *Analytical Biochemistry*, **1998**, 264, 98–110.

OIV (International Organization of Vine and Wine) *International code of oenological practices*, Paris, France, 2019.

Oliveira, C. M.; Ferreira, A. C. S.; Freitas, V. de; Silva, A. M. S. Oxidation mechanisms occurring in wines. *Food Research International*, **2011**, 44, 1115–1126.

Ooi, C. E.; Rabinovich, E.; Dancis, A.; Bonifacino, J. S.; Klausner, R. D. Copper-dependent degradation of the *Saccharomyces cerevisiae* plasma membrane copper transporter Ctr1p in the apparent absence of endocytosis. *The EMBO Journal*, **1996**, 15, 3515–3523.

Papadopoulou, D.; Roussis, I. G. Inhibition of the decline of linalool and alpha-terpineol in muscat wines by glutathione and n-acetyl cysteine. *Italian Journal of Food Science*, **2001**, 13, 413–419.

Park, S. K.; Boulton, R. B.; Noble, A. C. Automated HPLC analysis of glutathione and thiol-containing compounds in grape juice and wine using pre-column derivatization with fluorescence detection. *Food Chemistry*, **2000a**, 68, 475–480.

Park, S. K.; Boulton, R. B.; Noble, A. C. Formation of hydrogen sulfide and glutathione during fermentation of white grape musts. *American Journal of Enology and Viticulture*, **2000b**, 51, 91–97.

Pedneault, K.; Provost, C. Fungus resistant grape varieties as a suitable alternative for organic wine production: Benefits, limits, and challenges. *Scientia Horticulturae*, **2016**, 208, 57–77.

Penninckx, M. A short review on the role of glutathione in the response of yeasts to nutritional, environmental, and oxidative stresses. *Enzyme and Microbial Technology*, **2000**, 26, 737–742.

Penninckx, M. J. An overview on glutathione in *Saccharomyces* versus non-conventional yeasts. *FEMS Yeast Research*, **2002**, 2, 295–305.

Pérez-Serradilla, J. A.; Castro, M. L. de Role of lees in wine production: A review. *Food Chemistry*, **2008**, 111, 447–456.

Pérez-Torrado, R.; Barrio, E.; Querol, A. Alternative yeasts for winemaking: *Saccharomyces* non-*cerevisiae* and its hybrids. *Critical Reviews in Food Science & Nutrition*, **2018**, 58, 1780–1790.

Perino, A.; Vercesi, A.; Fregoni, M. Confronto tra diverse metodiche utilizzate nella determinazione dell'infezione da *Botrytis cinerea* sui mosti. *Vignevini*, **1994**, 7, 50–57.

Połeć-Pawlak, K.; Ruzik, R.; Lipiec, E. Investigation of Cd (II), Pb (II) and Cu (I) complexation by glutathione and its component amino acids by ESI-MS and size exclusion chromatography coupled to ICP-MS and ESI-MS. *Talanta*, **2007**, 72, 1564–1572.

Pozo-Bayón, M. Á.; Andújar-Ortiz, I.; Moreno-Arribas, M. V. Scientific evidences beyond the application of inactive dry yeast preparations in winemaking. *Food Research International*, **2009**, 42, 754–761.

Prillinger, H.; Esser, K. The phenoloxidases of the ascomycete *Podospora anserina*. *Molecular and General Genetics MGG*, **1977**, 156, 333–345.

Provenzano, M. R.; El Bilali, H.; Simeone, V.; Baser, N.; Mondelli, D.; Cesari, G. Copper contents in grapes and wines from a Mediterranean organic vineyard. *Food Chemistry*, **2010**, 122, 1338–1343.

Rao, Y.; Xiang, B.; Bramanti, E.; D'Ulivo, A.; Mester, Z. Determination of thiols in yeast by HPLC coupled with LTQ-Orbitrap mass spectrometry after derivatization with p-(Hydroxymercuri)benzoate. *Journal of Agricultural and Food Chemistry*, **2010**, 58, 1462–1468.

Ribéreau-Gayon, J.; Riberau-Gayon, P.; Seguin, G. Botrytis cinerea in enology. In *The biology of Botrytis*; Coley-Smith, J. R.; Verhoeff, K.; Jarvis, W. R., Eds.; Academic Press: London, UK, 1980, 251–274.

Ribéreau-Gayon, P.; Dubourdieu, D.; Donèche, B.; Lonvaud, A. *Handbook of enology, The microbiology of wine and vinifications*, edition no. 2; John Wiley & Sons: West Sussex, UK, 2006. Vol. 1.

Salgues, M.; Cheynier, V.; Gunata, Z.; Wylde, R. Oxidation of grape juice 2-S-glutathionyl caffeoyl tartaric acid by *Botrytis cinerea* laccase and characterization of a new substance: 2, 5-di-S-glutathionyl caffeoyl tartaric acid. *Journal of Food Science*, **1986**, 51, 1191–1194.

Santos, M. C.; Nunes, C.; Saraiva, J. A.; Coimbra, M. A. Chemical and physical methodologies for the replacement/reduction of sulfur dioxide use during winemaking: review of their potentialities and limitations. *European Food Research and Technology*, **2012**, 234, 1–12.

Shanmuganathan, A.; Avery, S. V.; Willetts, S. A.; Houghton, J. E. Copper-induced oxidative stress in *Saccharomyces cerevisiae* targets enzymes of the glycolytic pathway. *FEBS Letters*, **2004**, 556, 253–259.

Singleton, V. L. Oxygen with phenols and related reactions in musts, wines, and model systems: observations and practical implications. *American Journal of Enology and Viticulture*, **1987**, 38, 69–77.

Singleton, V. L.; Salgues, M.; Zaya, J.; Trousdale, E. Caftaric acid disappearance and conversion to products of enzymic oxidation in grape must and wine. *American Journal of Enology and Viticulture*, **1985**, 36, 50–56.

Smidt, O. de; Du Preez, J. C.; Albertyn, J. The alcohol dehydrogenases of *Saccharomyces cerevisiae*: a comprehensive review. *FEMS Yeast Research*, **2008**, 8, 967–978.

Soleas, G. J.; Diamandis, E. P.; Goldberg, D. M. Wine as a biological fluid: history, production, and role in disease prevention. *Journal of Clinical Laboratory Analysis*, **1997**, 11, 287–313.

Sonni, F.; Clark, A. C.; Prenzler, P. D.; Riponi, C.; Scollary, G. R. Antioxidant action of glutathione and the ascorbic acid/glutathione pair in a model white wine. *Journal of Agricultural and Food Chemistry*, **2011a**, 59, 3940–3949.

Sonni, F.; Moore, E. G.; Clark, A. C.; Chinnici, F.; Riponi, C.; Scollary, G. R. Impact of glutathione on the formation of methylmethine-and carboxymethine-bridged (+)-catechin dimers in a model wine system. *Journal of Agricultural and Food Chemistry*, **2011b**, 59, 7410–7418.

Steel, C. C.; Blackman, J. W.; Schmidtke, L. M. Grapevine bunch rots: impacts on wine composition, quality, and potential procedures for the removal of wine faults. *Journal of Agricultural and Food Chemistry*, **2013**, 61, 5189–5206.

Sun, X.; Liu, L.; Ma, T.; Yu, J.; Huang, W.; Fang, Y.; Zhan, J. Effect of high Cu2+ stress on fermentation performance and copper biosorption of Saccharomyces cerevisiae during wine fermentation. *Food Science and Technology*, **2019**, 39, 19–26.

Thakur, N. S.; Thakur, A.; Joshi, V. K.; Sharma, S. K. Botrytized wines: A review. *International Journal of Food and Fermentation Technology*, **2018**, 8, 1–13.

Thurston, C. F. The structure and function of fungal laccases. *Microbiology*, **1994**, 140, 19–26.

Tirelli, A.; Fracassetti, D.; Noni, I. de Determination of reduced cysteine in oenological cell wall fractions of *Saccharomyces cerevisiae*. *Journal of Agricultural and Food Chemistry*, **2010**, 58, 4565–4570.

Ugliano, M. Enzymes in winemaking. In *Wine chemistry and biochemistry*; Moreno-Arribas, M. V.; Polo, M. C., Eds.; Springer: New York, USA, 2009, 103–126.

Ugliano, M.; Kwiatkowski, M.; Vidal, S.; Capone, D.; Siebert, T.; Dieval, J.-B.; Aagaard, O.; Waters, E. J. Evolution of 3-mercaptohexanol, hydrogen sulfide, and methyl mercaptan during bottle storage of Sauvignon blanc wines. Effect of glutathione, copper, oxygen exposure, and closure-derived oxygen. *Journal of Agricultural and Food Chemistry*, **2011**, 59, 2564–2572.

Vanni, A.; Anfossi, L.; Pessione, E.; Giovannoli, C. Catalytic and spectroscopic characterisation of a copper-substituted alcohol dehydrogenase from yeast. *International Journal of Biological Macromolecules*, **2002**, 30, 41–45.

Vest, K. E.; Zhu, X.; Cobine, P. A. Copper Disposition in Yeast. In *Clinical and Translational Perspectives on WILSON DISEASE*, edition 1; Kerkar, N.; Roberts, E. A., Eds.; Academic Press an imprint of Elsevier: Massachusetts, USA, 2019, 115–126.

Vortkamp, A.; Rossetto, M.; Vanzani, P.; Di Paolo, M. L.; Klärner, S.; Muno-Bender, J.; Schneider, I.; Schnell, S.; Rigo, A.; Rauhut, D. Characterisation of laccase activity of two strains of *Botrytis cinerea* . *Journal of Plant Pathology*, **2013**, 95, 63–68.

Vuilleumier, S. Bacterial glutathione S-transferases: what are they good for? *Journal of Bacteriology*, **1997**, 179, 1431–1441.

Wang, Z. *Food enzymology*. China Light Industry Press: China, Beijing, 1990.

Waterhouse, A. L. Wine phenolics. *Annals of the New York Academy of Sciences*, **2002**, 957, 21–36.

Wilker, K. L. How can I avoid oxygen exposure with a white must? In *Winemaking problems solved*, edition 1; Butzke C., Ed.; Woodhead Publishing Limited: Cambridge, UK, 2010a, 35–36.

Wilker, K. L. How should I treat a must from white grapes containing laccase? In *Winemaking problems solved*, edition 1; Butzke C., Ed.; Woodhead Publishing Limited: Cambridge, UK, 2010b, 33–34.

Winge; Nielson, K. B.; Gray, W. R.; Hamer, D. H. Yeast metallothionein. Sequence and metal-binding properties. *Journal of Biological Chemistry*, **1985**, 260, 14464–14470.

Yamaguchi-Iwai, Y.; Serpe, M.; Haile, D.; Yang, W.; Kosman, D. J.; Klausner, R. D.; Dancis, A. Homeostatic regulation of copper uptake in yeast via direct binding of MAC1 protein to upstream regulatory sequences of FRE1 and CTR1. *Journal of Biological Chemistry*, **1997**, 272, 17711–17718.

Zouari, N.; Romette, J.-L.; Thomas, D. Purification and properties of two laccase isoenzymes produced by *Botrytis cinerea*. *Applied Biochemistry and Biotechnology*, **1987**, 15, 213–225.

Chapter 2

Influence of glutathione on yeast fermentation efficiency under copper stress

Copper in grape musts can influence the fermentation efficiency of *Saccharomyces cerevisiae* during winemaking. The present study revealed the impact of glutathione addition on yeast strains with variable copper sensitivity. The antioxidant glutathione increased yeast vitality and fastened sugar metabolism at copper concentrations up to 0.39 mM. A significant accumulation of acetaldehyde at high copper concentrations was mitigated by the addition of 20 mg·L^{-1} glutathione. Low recovery of glutathione added implicated a complexation of both compounds. Specific alcohol dehydrogenase (ADH) activity was inhibited or reduced in the enzyme extracts of the copper-stressed yeast cells. The activity was restored in fermentations with glutathione at a copper concentration of 0.16 mM. At low copper concentrations, glutathione decreased ADH activity presumably due to complexation of essential copper amounts. Results provide important information on the use of glutathione as an antioxidant in winemaking to counteract negative effects of copper-rich musts on copper-sensitive yeast strains.

Keywords: glutathione, copper stress, *Saccharomyces cerevisiae*, fermentation efficiency, specific alcohol dehydrogenase activity

---

1 Introduction

Copper can be found in grape must and wine for numerous reasons. These are the use of copper-based fungicides that accumulate in vineyard soils and, thus, in the grapevine,[1] the migration from certain materials of the winemaking equipment such as brass or the addition of copper sulfate or citrate to reduce sulfuric off-odors.[2] The narrow optimum concentration of copper leads to its ambivalence of essentiality and toxicity for *Saccharomyces cerevisiae*. At low concentrations, it is an essential micronutrient, a cofactor for enzymes, and involved in electron transport and redox reactions.[3] Excess copper amounts create oxidative stress in *S. cerevisiae* that reduces yeast vitality by impairing membrane integrity[4] and affecting enzyme activities.[5] In winemaking, this can have negative effects on fermentation efficiency of the yeast and eventually affects wine quality, e.g., in terms of volatile acid.[6,7] The influence of copper concentrations between 0.15 and 0.30 mM on different fermentation parameters has already been published.[8] It has been shown that copper compromised fermentation efficiency but the extent depends on the copper concentration and the yeast strain used.

The tripeptide glutathione (GSH) provides a prospective solution to reduce any negative effects of copper on fermentation. GSH can be used by the yeasts as an important cellular defense mechanism against heavy metal stress and oxidative damage.[9,10] It is synthesized intracellularly in two steps by γ-glutamylcysteine synthetase (GSH1) and GSH synthetase (GSH2)[9] or can be assimilated from the extracellular environment by specific transporters.[11,12] It acts as a strong cellular redox buffer[13] and protects cells by scavenging free radicals via its thiol moiety,[14] the reduction of peroxides by glutathione peroxidase, or the complexation of heavy metals, catalyzed by glutathione S-transferases.[15]

The use of GSH as an antioxidant in must and wine has recently received attention and its application is discussed controversially.[16] Its inhibitory property against oxidative browning is well established,[17] and the influence of GSH addition on the stability of aroma and color of wine has been reported.[18,19] However, GSH might play a role in the formation of sulfuric off-flavors[16] or the discoloration of white wines due to the formation of colored compounds.[20] A lack of studies exists on the complex interactions of yeast, copper stress, and GSH.

This study focuses on the influence of copper-induced stress on two different *S. cerevisiae* strains during fermentation of white grape must and the mitigating effect of GSH addition. The results provide new insights into the interaction of GSH and copper during fermentation and reveal the effect on different fermentation parameters like acetaldehyde accumulation, specific alcohol dehydrogenase (ADH) activity, sugar degradation, and yeast vitality.

2 Materials and methods

Yeast strains and chemicals. Two industrial *S. cerevisiae* strains Oenoferm X-treme (yeast 1) (Erbslöh, Geisenheim, Germany) and Anaferm Riesling (yeast 2) (Zefüg, Alzey, Germany) were chosen for microvinifications (MV) based on preliminary experiments. Copper-(II) sulfate pentahydrate ($CuSO_4 \cdot 5H_2O$) was acquired from AppliChem (Darmstadt, Germany) and reduced L-glutathione ≥98% was from Sigma-Aldrich (Steinheim, Germany). Potassium dihydrogen phosphate ≥99% (Carl Roth, Karlsruhe, Germany), disodium hydrogen phosphate dihydrate (neoFroxx, Einhausen, Germany), and phosphate-buffered saline (PBS) (Roti-Cell 10x PBS, Carl Roth, Karlsruhe, Germany) were used for cell suspensions. Fluorescein diacetate (FDA) 97% was purchased from Alfa Aesar (Ward Hill, MA). Standards for quantification were acetaldehyde ≥99.5% (ACROS Organics, Geel, Belgium), D(-)-fructose >99.5%, and D(+)-glucose monohydrate both acquired from Carl Roth (Karlsruhe, Germany). The derivatization reagents for GSH analysis *ortho*-phthaldialdehyde (OPA) and 2-aminoethanol >98%, the buffer component anhydrous borax >99% were all from Sigma-Aldrich (Steinheim, Germany). Eluents consisted of acetonitrile ≥99.9%, high-performance liquid chromatography-liquid chromatography-mass spectroscopy (HPLC-LC-MS) grade (VWR Chemicals, Darmstadt, Germany). Bovine serum albumin (BSA), Coomassie Brilliant Blue G-250, and *ortho*-phosphoric acid 85% were obtained from Merck (Darmstadt, Germany). Further reagents for the ADH activity assay were sodium pyrophosphate decahydrate ≥99.5% (ACROS Organics, Geel, Belgium), nicotinamide adenine dinucleotide (NAD)

(AppliChem, Darmstadt, Germany), and semicarbazide hydrochloride 99% (Thermo Fisher, Kandel, Germany).

Microvinifications. Microvinifications (MV) were carried out in sterile 2.5 L brown glass bottles closed with a cap containing a septum. The bottles were equipped with a fermentation lock and a sample tubing connected to a three-way valve with a safeflow membrane (Discofix C with a safeflow valve and 50 cm tubing, Braun Melsungen, Melsungen, Germany). A magnetic stir bar was placed in each bottle that was filled with 2 L sterilized Riesling grape must kindly provided by the Institute of Viticulture and Enology, DLR Rheinpfalz (Neustadt/Weinstrasse, Germany). Must sterilization was conducted at 80 °C for 60 s. Prior to that bentonite was added (1 $g \cdot L^{-1}$) and the must was filtered to prevent protein precipitation during the thermal treatment. No SO_2 was added to the must to exclude any additional antioxidant activities besides GSH. The musts were spiked with copper sulfate to obtain three different Cu^{2+} concentrations: 2.5 $mg \cdot L^{-1}$ (0.04 mM), 10 $mg \cdot L^{-1}$ (0.16 mM), and 25 $mg \cdot L^{-1}$ (0.39 mM). The original copper content of the must was 1.2 $mg \cdot L^{-1}$ (0.02 mM) and was adjusted by adding $CuSO_4 \cdot 5H_2O$ before fermentation. Each must was fermented in triplicate without GSH addition (MV1–MV4) or with GSH addition (MV1-GSH to MV4-GSH) of 20 $mg \cdot L^{-1}$ (0.07 mM) after 24 h of fermentation. GSH was added after 24 h to provide enough time for the adaption of the yeasts to the new environment after rehydration, to ensure that fermentation has started under the different conditions, and a high concentration of active yeast cells was available. Musts were inoculated with 0.2 $g \cdot L^{-1}$ of yeast 1 or yeast 2 on day 0 according to the manufacturer's instructions previously rehydrated with 0.3 $g \cdot L^{-1}$ of the mineral and vitamin supplement GoFerm (Lallemand). GSH (MV1-GSH to MV4-GSH) and 0.3 $g \cdot L^{-1}$ of the complex nutrient supplement Fermaid E (Lallemand) (MV1−MV4 and MV1-GSH to MV4-GSH) were added to the must via the three-way valve. Both compounds were suspended in 1 mL of sterile water. Preliminary experiments showed a content of 0.5 mg of $GSH \cdot g^{-1}$ Fermaid. This content was subtracted from the GSH amount added to maintain the GSH concentration of 20 $mg \cdot L^{-1}$ in MV1-GSH to MV4-GSH. After yeast inoculation, microvinifications were kept for 2 h at room temperature and fermentation was continued at 16 ± 1 °C.

Sample preparation. Sampling from MVs was carried out during 28 days of fermentation. The bottles were placed on a magnetic stirrer for 1 min, and a volume of 40 mL (day 1–3) or 30 mL (day 4–28) was taken with a syringe from each bottle over the sterilized safeflow membrane. Yeast cells were harvested by centrifugation (9000 *g*, 3 min). The supernatant was stored at -18 °C until further analysis of sugars, acetaldehyde, GSH, and copper. Before analysis, samples were filtered through 0.2 µm Chromafil RC-20/15 MS filters (Macherey-Nagel, Düren, Germany). The cell pellet was washed twice with 20 mL of 0.1 M phosphate buffer, pH 7.4, and was finally resuspended with 30 mL of the same buffer. A volume of 1 mL cell suspension was taken for direct flow cytometry (FCM) analysis. The remainder was frozen at -18 °C for cell disruption and preparation of cell extracts.

Flow cytometry analysis. FCM was used to monitor cell vitality throughout the fermentation with the cell-permeable esterase substrate fluorescein diacetate (FDA). A modified method by Panton and Jones was used.[21] Cell suspensions were diluted with 1x PBS buffer to approximately 10^6 cells·mL^{-1}. An aliquot of 990 µL of the cell suspension was incubated with 10 µL of 0.5 mg·mL^{-1} of FDA in acetone for 3 min at room temperature in the dark. The stained samples were homogenized and subjected to FCM analysis. The analysis was performed using a CyFlow Cube 6 (Sysmex Partec, Germany) equipped with a 488 nm blue laser, a forward scatter detector (FSC), side scatter detector (SSC), and four fluorescence detectors. For fluorescence measurements, only the FL1 detector (536 ± 20 nm) was triggered. Raw data files were processed using the software FCS Express 4 (De Novo Software). The yeast cell population was identified and selected by gating in an FSC/SSC density plot. A histogram of the FL1 fluorescence was obtained for the gated region. It was considered for the differentiation of FDA positive (metabolic activity) and negative (metabolic inactivity) stained cells. The percentage of positively stained cells presented cell vitality. The signal intensity of FDA negative cells was identified by measuring an unstained sample of yeast cell suspension. This low fluorescence can be attributed to the autofluorescence of the cells.

Preparation of yeast cell-free extracts for enzyme activity assay. Ultrasound disruption of yeast cells was performed using a Hielscher UP200St ultrasonic processor equipped with the sonotrode S26d7 (Hielscher, Teltow, Germany). The tip of the sonotrode with a diameter of 7 mm was located in the center of the cell suspension. Sonication was conducted for 15 min at 80% amplitude and a duty cycle of 50% (on/off-time 0.5/0.5 s). The samples were cooled in an ice bath during treatment to keep the temperature below 33 °C. The cell extract was centrifuged at 11000 *g* for 10 min and the supernatant was immediately used for the analysis of specific enzyme activity described below. The protein content was determined using the Bradford assay[22] and BSA as the protein standard. The protein stock solution (1 $mg \cdot mL^{-1}$) was dissolved in different concentrations in 0.1 M phosphate buffer (pH 7.4). For calibration, the absorption was measured with a FLUOstar Omega microplate reader spectrophotometer (BMG Labtech, Ortenberg, Germany) at 595 nm and compared with the cell extracts. The residual cell extract was stored at -18 °C for the quantification of intracellular copper content.

ADH activity assay. ADH activity was measured spectrophotometrically according to a method of Mauricio et al. with slight modifications.[23] Reaction mixtures contained 1.25% (v/v) cell extract, 0.1 M sodium pyrophosphate buffer (pH 9), 70 mM semicarbazide, and 2 mM NAD. Ethanol was added as the substrate in a concentration of 500 mM to start the reaction. Extinction was measured at 340 nm with a Fluostar Omega spectrophotometer at 25 °C for 5 min (intervals of 30 s). Each sample was prepared in triplicate. Enzyme activity units (U) are expressed as µmol of NADH formed per minute ($\varepsilon_{340} = 6.3 \times 10^3$ L $(cm \cdot mol)^{-1}$).[24]

HPLC sugar analysis. The concentrations of fructose and glucose in the samples were determined using a Smartline HPLC system (Knauer, Berlin, Germany) equipped with two pumps, RI detector 2300, column oven, autosampler (Midas, Spark Holland, Emmen, Netherlands) and a Nucleodur 100-5NH2-RP column (150 x 4.6 mm, particle size 5 µm) from Macherey-Nagel (Düren, Germany). Isocratic elution with acetonitril/water (85:15, v/v) at 40 °C with a constant flow of 1.0 $mL \cdot min^{-1}$ was used for analysis. The injection volume was 10 µL. For the quantification of sugars, standard solutions of glucose and fructose in different concentrations were used for external calibration.

Evaluation of fermentation progress. The sum of the glucose and fructose concentration was expressed as total sugar content. The total sugar content on day 0 was used to calculate the percentage of sugar degradation during fermentation for each

MV. The resulting sigmoidal curve was fitted by a modified Gompertz equation described by Zwietering et al.[25] using the software OriginPro 8G (OriginLab, Northampton, MA). The model describes the acceleration from a start value of zero to a maximal value V_{max} (maximum rate of sugar degradation) in a specific period of time that results in a lag phase λ. The final phase that determines the asymptote was neglected since each fermentation progress is expected to reach nearly 100% sugar degradation.

Table 2.1: Lag phase λ of sugar degradation of the yeast Oenoferm X-treme (yeast 1) and Anaferm Riesling (yeast 2) in microvinifications (MV) with different copper concentrations[a,b]

		yeast 1		yeast 2	
		-	GSH	-	GSH
MV	Cu (mM)	λ (d)	λ (d)	λ (d)	λ (d)
1	control	1.79 ± 0.13a	1.22 ± 0.03a	1.87 ± 0.09a	1.33 ± 0.08a
2	0.04	1.61 ± 0.05a	1.15 ± 0.10a	1.76 ± 0.07a	1.29 ± 0.10a
3	0.16	2.40 ± 0.14b	1.50 ± 0.14a,b	1.97 ± 0.20a	1.41 ± 0.04a
4	0.39	3.80 ± 0.37c	1.69 ± 0.21b	1.87 ± 0.40a	1.32 ± 0.16a

[a]Values are calculated by modelling the percentage of sugar degradation with a modified Gompertz equation. [b]Different letters indicate significant differences within a column

Table 2.2: Maximum rate of sugar degradation V_{max} of the yeast Oenoferm X-treme (yeast 1) and Anaferm Riesling (yeast 2) in microvinifications (MV) with different copper concentrations[a,b]

		yeast 1		yeast 2	
		-	GSH	-	GSH
MV	Cu (mM)	V_{max} (d^{-1})	V_{max} (d^{-1})	V_{max} (d^{-1})	V_{max} (d^{-1})
1	control	0.17 ± 0.01a	0.20 ± 0.00a	0.15 ± 0.00a	0.18 ± 0.01a
2	0.04	0.16 ± 0.00a	0.19 ± 0.00a	0.15 ± 0.01a	0.19 ± 0.00a
3	0.16	0.13 ± 0.00b	0.18 ± 0.01a	0.15 ± 0.00a	0.19 ± 0.01a
4	0.39	0.11 ± 0.01b	0.14 ± 0.01b	0.15 ± 0.01a	0.18 ± 0.01a

[a]Values are calculated by modelling the percentage of sugar degradation with a modified Gompertz equation. [b]Different letters indicate significant differences within a column.

Ultra-HPLC GSH analysis with precolumn derivatization. A modified method of an automated precolumn derivatization of thiols was used for the quantification of GSH.[26] Samples were analyzed with a Prominence UFLC system (Shimadzu, Kyoto, Japan) equipped with a Prominence DGU-20A 5R degasser, two Nexera X2 LC-30AD pumps, a Nexera SIL-30AC Prominence autosampler kept at 4 °C, an Acquity UPLC HSS T3-

column (150 x 2.1 mm, particle size 1.8 µm) from Waters (Milford, MA), a CTO-20AC Prominence column oven kept at 40 °C, and a Shimadzu RF-10A XL fluorescence detector. A volume of 50 µL of the sample was filled into a flat bottom insert and the following derivatization procedure was used: 20 µL of OPA (4 mg·mL^{-1} in acetone) and 1 µL of the air were aspirated into the sample loop, the needle was dipped into the washing port, 20 µL of 2-aminoethanol (4 µL in 1 mL 0.2 M borate buffer, pH 7.4) was aspirated, the needle was dipped into the washing port, the sample loop with both reagents was drained into the sample vial, and the reaction mixture was agitated. After a reaction time of 1 min, 2 µL was injected. Analysis was conducted with the following gradient program at a flow rate of 0.4 mL·min^{-1} using eluent A 50 mM sodium acetate buffer (pH 5.7) and eluent B acetonitril/water (95:5, v/v): 0 min, 5% B; 20 min, 25% B; 21 min, 100% B; 25 min, 100% B; 26 min, 5% B; 30 min, 5% B. The GSH standard solution for quantification was prepared in 5 mM sodium acetate buffer (pH 4).

Gas chromatography (GC) analysis. Acetaldehyde was analyzed by Headspace-GC-FID (Agilent 6890 Series, SIM, Ratingen, Germany) using a Combi-PAL-autosampler (CTC Analytics, Zwingen, Switzerland) and an OPTIMA WAXplus capillary column (30 m x 250 µm x 0.25 µm, Macherey-Nagel, Düren, Germany). Headspace vials (20 mL) were filled with 1 mL of the sample. Samples were incubated at 60 °C for 5 min before the injection of 500 µL. The inlet temperature was set at 200 °C and the carrier gas was nitrogen with a constant flow of 0.7 mL·min^{-1}. The following oven temperature program was used: 40 °C (1 min); temperature increase, 5 °C·min^{-1}; final temperature 50 °C (10 min). The detector gases H_2 (40 mL·min^{-1}) and air (450 mL·min^{-1}) were applied, and the detector temperature was set at 250 °C. Standard solutions of acetaldehyde in different concentrations were used for quantification by external calibration. Identification of acetaldehyde on two different columns showed RI of 681 on the OPTIMA WAXplus column and 496 on an Rtx-5 w/Integra Guard (Crossbond 5% diphenyl–95% dimethyl polysiloxane, 30 m x 250 µm x 0.25 µm) capillary column (Restek, Bellefonte, PA), respectively.

Analysis of metal content. Copper was determined in the supernatant as extracellular copper content (Cu_{ex}) and in the cell extracts as intracellular copper content (Cu_{in}). The analysis was performed with an AA240Z Zeeman atomic absorption spectrometer

(Varian, Palo Alto, CA) equipped with a PSD 120 autosampler and an AA 240FS flame according to the method L 00.00-19/2 of the official collection of test methods pursuant to § 64 LFGB, § 38 TabakerzG, § 28b GenTG.[27]

Statistical analysis. Statistical analysis of the results was conducted with XLSTAT (Version 2014.4.06, AddinSoft Technologies, Paris, France). In the case of variance homogeneity, an analysis of variance followed by a Tukey test was used and for pairwise comparisons a Student's *t*-test both with a selected significance level of $p < 0.5$.

3 Results and discussion

Fermentation progress. The parameters yeast vitality and maximum rate of sugar degradation V_{max} were used to compare the overall fermentation performance of the different MV trials. The adaption phase of the yeast at the beginning of fermentation was included by determining the lag phase λ of sugar degradation. Sugar degradation allows direct conclusions on the fermentation progress, whereas yeast vitality is a parameter that reflects fermentation performance and stress conditions via metabolic activity, i.e., esterase activity[28] that still occurs when fermentation is completed.

Tables 2.1 and **2.2** show the results of sugar degradation for both yeasts. Although yeast 2 was not affected by copper addition, yeast 1 presented significantly longer lag phases at high copper concentrations (MV3 and MV4). The lag phase of MV4 was approximately 2 days longer than in MV1 and MV2. The variants with low copper concentrations (MV1 and MV2) did not differ considerably but MV2 showed a slightly shorter lag phase. This indicates that a low copper concentration of 0.04 mM might promote sugar degradation in the early fermentation stage. It has been shown that a copper concentration of 0.5 mM improved growth activity, but no effect on fermentation efficiency has been observed.[29]

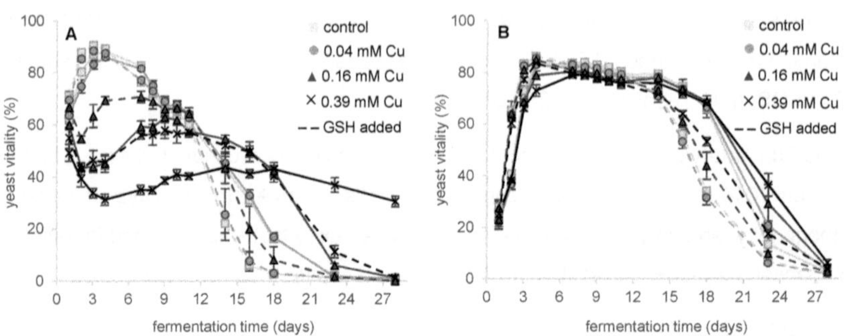

Figure 2.1: Yeast vitality in microvinifications (MV) with different copper concentrations. Dashed lines represent MVs with an addition of 20 mg·L^{-1} GSH. (A) Oenoferm X-treme (yeast 1); (B) Anaferm Riesling (yeast 2).

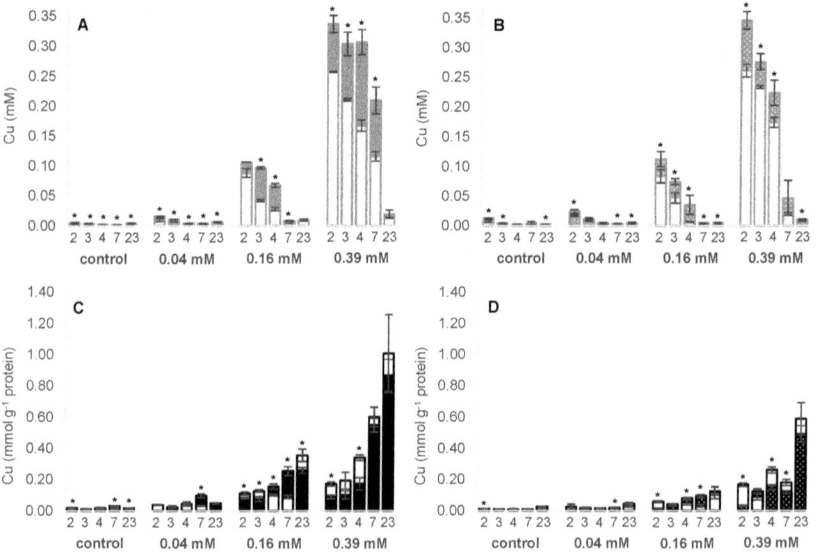

Figure 2.2: Extracellular copper (Cu_{ex}) and intracellular copper (Cu_{in}) content in microvinifications (MV) with different copper concentrations. Bars are overlain (mean ± standard deviation (SD)) and white bars represent MVs with an addition of 20 mg·L^{-1} GSH. (A) Cu_{ex} Oenoferm X-treme (yeast 1), (B) Cu_{ex} Anaferm Riesling (yeast 2), (C) Cu_{in} yeast 1, (D) Cu_{in} yeast 2. * indicates significant differences between MV trials with or without GSH addition.

GSH addition to copper-stressed yeast cells led to a shortening of the lag time. The difference of 2 days was reduced to approximately 0.5 days between MV4-GSH and the low copper trials MV1-GSH and MV2-GSH. Values of V_{max} were inversely proportional to the copper concentration, indicating a delayed fermentation under high copper stress. This is described as a stuck and sluggish fermentation, being a result of various stress conditions.[30] The effect can be reduced by GSH addition in MVs with a copper concentration up to 0.16 mM. An initially decelerated sugar degradation has already been described for copper concentrations from 0.15 to 0.30 mM resulting in a stuck and sluggish fermentation.[8] The values of λ and V_{max} for yeast 2 did not differ significantly. This might implicate that the threshold of copper leading to stress conditions for yeast 2 is higher. Previous studies already observed a difference in copper resistance or sensitivity of different yeast strains.[6–8,31] Copper thresholds, which still ensured a good reducing sugar utilization, also differed between yeast strains.[8] At the end of fermentation, all trials had similar residual sugar levels with a slight increase proportional to the copper concentrations (**Table S 2.1**). The sole exception is MV4 with yeast 1 with an approximately five-fold higher final sugar level compared to MV4-GSH. This again can be explained by the high copper sensitivity of yeast 1 compared to yeast 2 and is further proof for the possibility of the mitigation of a copper-stressed-induced stuck and sluggish fermentation by GSH addition.

Yeast vitality confirmed the higher copper sensitivity of yeast 1 compared to yeast 2 (**Figure 2.1**). Copper exposure up to 0.04 mM (MV1) for yeast 1 and up to 0.39 mM (MV4) for yeast 2 did not cause stress conditions. Under these conditions, a comparable

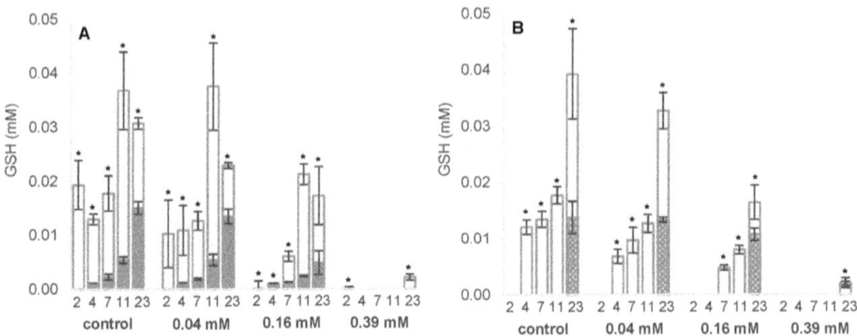

Figure 2.3: GSH concentrations in microvinifications (MV) with different copper concentrations. Bars are overlain (mean ± SD) and white bars represent MVs with an addition of 20 mg·L^{-1} GSH. (A) Oenoferm X-treme (yeast 1), (B) Anaferm Riesling (yeast 2). * indicates significant differences between MV trials with or without GSH addition.

progress of yeast vitality with three phases was observed: a short exponential and stationary phase followed by a long phase of decreasing yeast vitality. In contrast, yeast 1 exposed to 0.16 and 0.39 mM copper (MV3 and MV4) showed a first phase of decreasing vitality followed by the three phases mentioned above. The decrease has also been observed in a previous study for a very high copper concentration of 1.50 mM in grape must.[29] It has been described as a period of adaption to the stress conditions where proliferation was restrained. Other studies described an initial growth inhibition of yeasts in a 0.25 mM copper-enriched must[7] and in 0.5–1.5 mM copper-enriched model grape juices.[31] The highest copper stress led to the longest adaption phase of 4 days for MV4 compared to the other MV trials. This correlates with the longest lag phase of sugar degradation with nearly 4 days for MV4. The other three phases of MV4 were less pronounced with a low maximum increase in vitality and only a slight decrease until day 28. All other variants already reached a low minimum vitality up to 5–10 days earlier. This can be explained by a faster nutrient depletion by the more vital cells in contrast to MV4. A lower vitality of the yeast cells and the corresponding lower metabolism result in a longer availability of nutrients and, thus, a longer maintenance of the (low) vitality in MV4. GSH addition changed the development of yeast vitality to clear phases in MV4-GSH with a significantly higher maximum vitality than in MV4 and a significantly lower minimum vitality after day 28. A significantly increased maximum yeast vitality was also observed for MV3-GSH in contrast to MV3. The effect of 0.04 mM copper on vitality in MV2 and MV2-GSH was negligible, showing almost the same progress of phases as the control.

Fermentation performance of yeast 1 was already affected by a low copper concentration (0.16 mM) that lies within the permitted range in the European Union and South Africa (≤20 mg·L^{-1}, 0.31 mM).[7] It has been suggested that this legal limit should be lowered to a threshold of 0.20 mM because of the inhibited fermentation performance.[8] These authors described that cell growth has not been affected, but the present study revealed a pronounced influence on yeast vitality. Apparently, determination of yeast vitality is a more sensitive indicator for fermentation performance, because it responds more clearly to stress conditions.

Copper concentrations. Copper concentrations were analyzed for both yeasts in the must (Cu_{ex}, **Figure 2.2 A, B**) and intracellularly in the cell extracts (Cu_{in}, **Figure 2.2 C, D**). Cu_{in} values were corrected for the amount of protein in the extract and, thus, describe the average copper content inside the yeast cell. Initial Cu_{in} concentrations of the rehydrated yeast cells were 0.016 mmol·g^{-1} for yeast 1 and 0.001 mmol·g^{-1} for yeast 2. Cu_{ex} concentrations for both yeasts showed a strong decrease during fermentation, ranging from not detectable to 0.02 mM after day 23 for all MV trials. Yeast cells have the ability to complex and import copper into the cell to maintain copper homeostasis.[32] Under high copper exposition, two phases of biosorption have been described. Fast extracellular adsorption to the cell surface is followed by a slow intracellular accumulation.[8,33] This agrees with a slower increase in Cu_{in} concentrations in contrast to the fast decrease in Cu_{ex} concentrations during fermentation. The remaining copper obviously was adsorbed to cell debris (extracellular copper) that was removed from the cell extracts prior to analysis. Several studies reported the removal of copper from the fermentation medium by biosorption of the yeast.[8,29,31,34] This was particularly visible for MVs with high copper concentrations (MV3 and MV4). For yeast 1, a slower initial decrease in Cu_{ex} concentrations was observed in MV3 and MV4 compared to MV3-GSH and MV4-GSH. This might be a reason for the higher stressed yeast cells without GSH addition, which resulted in a longer adaption phase and, thus, restrained proliferation. For yeast 2, Cu_{ex} values quickly decreased with and without GSH addition. However, yeast 2 generally showed lower Cu_{in} values than yeast 1. This implicates good extracellular copper adsorption of this yeast strain. The low copper uptake might be associated with the high copper resistance of yeast 2. Published results of copper biosorption are contradictory because some authors have described a higher removal rate by those strains that are more sensitive to copper,[8,31] whereas others have observed a better biosorption of the resistant strain.[34]

Although yeast 2 was more resistant to copper than yeast 1, an influence of GSH on Cu_{ex} and Cu_{in} contents was observed for both yeasts. In MV trials with high copper concentrations, considerably lower Cu_{ex} contents and higher Cu_{in} contents were observed when GSH was added in contrast to MVs without GSH addition. The reason for the inversed ratio from day 4 to 7 for MV3 and MV3-GSH cannot be explained, but it might be ascribed to the protection mechanisms against copper, which is mainly driven by the Cup1 metallothionein[32] that remained active in the cells in the presence of sufficient GSH. A relationship between copper detoxification and GSH is possible due to the metal-reducing property of GSH. The formation of copper(I) GSH complexes, which are then transported to metallothioneins, has been described.[35] A similar mechanism found for cadmium might be responsible for copper export from the cell via GSH S-conjugate export pumps.[36]

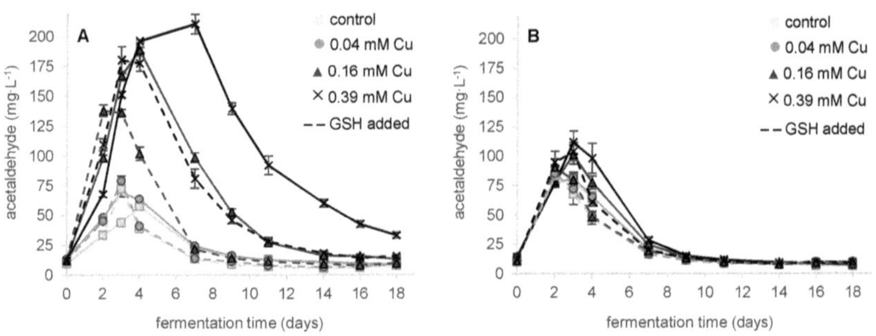

Figure 2.4: Acetaldehyde concentrations in microvinifications (MV) with different copper concentrations. Dashed lines represent MVs with an addition of 20 mg·L-1 GSH. (A) Oenoferm X-treme (yeast 1); (B) Anaferm Riesling (yeast 2).

Table 2.3: Specific ADH activity in the enzyme extract of the yeast Oenoferm X-treme (yeast 1) and Anaferm Riesling (yeast 2) in microvinifications (MV) with different copper concentrations[a,b]

		yeast 1		yeast 2	
		-	GSH	-	GSH
MV	Cu (mM)	specific activity ($U \cdot mg^{-1}$)	specific activity ($U \cdot mg^{-1}$)	specific activity ($U \cdot mg^{-1}$)	specific activity ($U \cdot mg^{-1}$)
1	control	6.44 ± 0.06a	2.98 ± 0.08a	3.12 ± 0.14a	3.23 ± 0.07a
2	0.04	5.26 ± 0.76a	3.29 ± 0.16a	2.86 ± 0.19a	3.22 ± 0.22a
3	0.16	0.00 ± 0.00b	1.60 ± 0.13b	2.73 ± 0.15a	3.94 ± 0.51a
4	0.39	0.00 ± 0.00b	0.00 ± 0.00c	0.00 ± 0.00b	0.00 ± 0.00b

[a]Determined using the ADH assay and corrected for the amount of protein in the extract. [b]Different letters indicate significant differences within a column.

GSH concentrations. After inoculation on day 0, no GSH was detected in the must of all MV trials. During fermentation, GSH concentrations generally increased and were obviously higher in MV with GSH addition (**Figure 2.3**). However, the recovery of GSH added (0.07 mM) on day 2 was very low, ranging from not detectable to 0.02 mM. It has been described that the oxidation of GSH is accelerated in the presence of copper, which may form complexes with glutathione disulfide (GSSG),[37] whereby the latter was not detected by the method used. This might explain the lower or not detectable GSH concentrations in MV with high copper concentrations. Yeast metabolism also affects GSH concentrations in the must. The role of GSH as an important stress response factor during sulfur and nitrogen deficiency, oxidative stress, or heavy metal detoxification[10] may govern the ratio of its assimilation or secretion by specific transporters.[11,12,36] It has already been described that GSH concentrations may increase[38,39] or decrease during fermentation.[39,40]

Acetaldehyde concentrations. Acetaldehyde accumulation in the must proportionally increased to the copper concentrations (**Figure 2.4**). The increase was more pronounced for the copper-sensitive yeast 1 than for yeast 2. Copper concentrations of ≥0.16 mM (MV3 and MV4) resulted in a significantly higher acetaldehyde accumulation, followed by a slow acetaldehyde decrease during fermentation. On day 18, MV3 and MV4 still showed significantly higher acetaldehyde concentrations than MV1 and MV2. The addition of GSH led to the mitigation of acetaldehyde accumulation and a faster acetaldehyde decrease. Results suggest that acetaldehyde is a sensitive indicator for copper-induced stress conditions for the yeast cells during fermentation. This was shown by its temporal accumulation due to copper stress, which can be reduced by GSH addition.

Acetaldehyde is known for its high reactivity and toxicity, but it appears that even the highest mean maximum concentration of 210.7 mg·L^{-1} in MV4 does not affect yeast fermentation efficiency. External addition of much higher acetaldehyde concentrations (≥400 mg·L^{-1}) significantly reduced cell population, glucose metabolism, and ethanol production.[41] In contrast, an addition of 580 mg·L^{-1} acetaldehyde even stimulated yeast cell growth under anaerobic conditions in the presence of 3–6% ethanol.[42,43] The underlying mechanism has not been clarified yet but might be connected to NAD^+ regeneration.[44] It has to be considered that high acetaldehyde concentrations are undesired in winemaking regarding the use of SO_2 as a preservative agent. Acetaldehyde is a strong binder of SO_2, reducing its functional properties in contrast to free SO_2.[44]

ADH activity. The method used in this experiment analyzed the specific ADH activity of the two isoenzymes ADHI and ADHII in the enzyme extract.[23] Differentiation can be made by thermal inactivation of ADHI,[45] but ADHII is found only in *S. cerevisiae* grown under aerobic conditions[46] and its expression is inhibited by glucose.[47] It is, therefore, assumed that only ADHI is expressed because fermentation was conducted under anaerobic conditions in a glucose-rich grape must.

Specific ADH activity was compared among different MVs after 1 week of fermentation. It was expected that the adaption phase was finished under high copper stress and MVs with low copper concentrations still showed a high yeast vitality. Results in **Table 2.3** show that ADH activity without GSH added decreases inversely proportional to copper concentrations. The impact is again more distinct for yeast 1 compared to yeast 2. No activity was detectable in MV3 and MV4 for yeast 1 and in MV4 for yeast 2. An *in vivo* ADH was certainly active because a decrease in acetaldehyde was observed in all MVs.

Two mechanisms are possibly responsible for the ADH inhibition by copper. It has been shown that the ADHI of *S. cerevisiae* is among the enzymes that are most prone to oxidation by copper.[5] This is caused by the copper-mediated Fenton reaction that promotes the formation of an elevated cellular level of free hydroxyl radicals ($OH^{\bullet}$). The second mechanism has been described as a replacement of zinc by copper or other transition metals, which has structural and catalytic functions in the ADH.[48] This displacement leads to reduced specific enzyme activity.[49,50] Quantification of intracellular zinc contents, however, did not show a decrease under high copper exposure (**Figure S 2.1**).

The influence of GSH and different copper concentrations on ADH activity was ambivalent. Regarding yeast 1, GSH addition led to a detectable ADH activity in MV3-GSH compared to MV3. This might be explained by the lowered copper exposition because GSH complexes extracellular copper. This can be proven with lower Cu_{in} values after day 7 with GSH added (MV3-GSH) compared to MV3. In contrast, the ADH activity in trials with low copper concentrations (MV1-GSH and MV2-GSH) was even lowered by the addition of GSH compared to MV1 and MV2. This might be explained by the essentiality of low amounts of copper for the cells and corresponding positive influence on ADH activity. The essential copper is complexed by GSH extracellularly, making it unavailable for enzymatic processes inside the cells. There was still a slightly higher activity in MV2-GSH than in MV1-GSH since more extracellular copper was added. This implies a balance between copper stress and copper deficiency that is apparently influenced by GSH. The difference between ADH activities in low copper trials, however, did not induce an effect on acetaldehyde accumulation. Even the low activities in MV1-GSH and MV2-GSH were within a range that apparently did not affect fermentation performance.

GSH acts as a substantial antioxidant and might give winemakers a new tool to handle copper-rich grape musts and reduce negative effects on fermentation efficiency. On the other hand, GSH addition may lead to copper deficiency, and, thus, the copper sensitivity of the yeast strain used has to be considered. Especially in regions where warm and wet climate requires the use of copper-based fungicides, GSH can provide a tool to achieve better wine quality. Further research is necessary to assess whether a positive impact on organoleptic wine properties by GSH addition to copper-rich musts can be expected. The use of GSH-enriched inactive dry yeast preparations in comparison to pure GSH also needs to be assessed.

Funding

This research project was financially supported by the German Ministry of Economics and Technology (via AiF) and the FEI (Forschungskreis der Ernährungsindustrie e.V., Bonn). Project AiF 18645N.

Acknowledgements

We thank Dr. Jürgen Fröhlich (Erbslöh, Geisenheim, Germany) for his supportive comments and Anja Stratmann (Institute of Nutritional and Food Sciences, Food Chemistry, University of Bonn) for performing the copper and zinc analysis.

Supporting information

Table S 2.1: Residual sugar levels at the end of fermentation for the yeast Oenoferm X-treme (yeast 1) and Anaferm Riesling (yeast 2) in microvinifications (MV) with different copper concentrations[a,b]

		yeast 1		yeast 2	
		-	GSH	-	GSH
MV	Cu (mM)	total sugar ($g \cdot L^{-1}$)	total sugar ($g \cdot L^{-1}$)	total sugar ($g \cdot L^{-1}$)	total sugar ($g \cdot L^{-1}$)
1	control	0.91 ± 0.19a	0.61 ± 0.17a	0.66 ± 0.22a	0.53 ± 0.30a
2	0.04	1.00 ± 0.20a	0.77 ± 0.14a	1.00 ± 0.18a	0.48 ± 0.11a
3	0.16	1.40 ± 0.02a	1.01 ± 0.41a	0.95 ± 0.31a	0.90 ± 0.38a
4	0.39	5.77 ± 0.22b	1.09 ± 0.17a	1.02 ± 0.20a	1.07 ± 0.29a

[a]Different letters indicate significant differences within a column. [b]Total sugar content was determined by the sum of glucose and fructose.

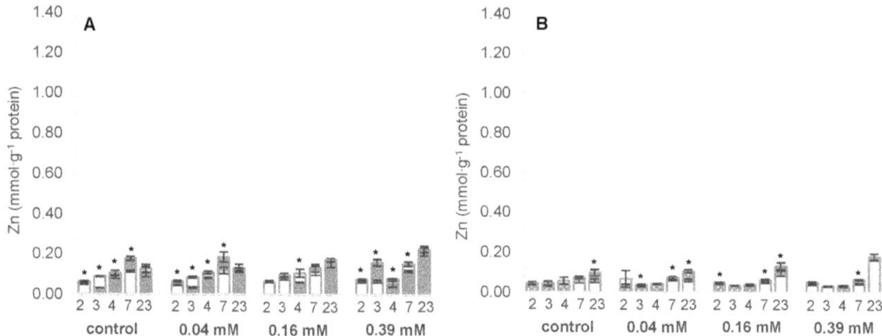

Figure S 2.1: Intracellular zinc contents (Zn_{in}) corrected for the amount of protein in the cell extracts in microvinifications (MV) with different copper concentrations. Bars are overlain (mean ± SD) and white bars represent MVs with an addition of 20 $mg \cdot L^{-1}$ of GSH; A: Zn_{in} Oenoferm X-treme (yeast 1), B: Zn_{in} Anaferm Riesling (yeast 2). * indicates significant differences between MV trials with or without GSH addition.

References

(1) Besnard, E.; Chenu, C.; Robert, M. Influence of organic amendments on copper distribution among particle-size and density fractions in Champagne vineyard soils. *Environ. Pollut.* **2001**, 112, 329−337.

(2) Clark, A. C.; Wilkes, E. N.; Scollary, G. R. Chemistry of copper in white wine: a review. *Aust. J. Grape Wine Res.* **2015**, 21, 339−350.

(3) Cervantes, C.; Gutierrez-Corona, F. Copper resistance mechanisms in bacteria and fungi. *FEMS Microbiol. Rev.* **1994**, 14, 121−137.

(4) Avery, S. V.; Howlett, N. G.; Radice, S. Copper toxicity towards *Saccharomyces cerevisiae*: dependence on plasma membrane fatty acid composition. *Appl. Environ. Microbiol.* **1996**, 62, 3960−3966.

(5) Shanmuganathan, A.; Avery, S. V.; Willetts, S. A.; Houghton, J. E. Copper-induced oxidative stress in *Saccharomyces cerevisiae* targets enzymes of the glycolytic pathway. *FEBS Lett.* **2004**, 556, 253−259.

(6) Cavazza, A.; Guzzon, R.; Malacarne, M.; Larcher, R. The influence of the copper content in grape must on alcoholic fermentation kinetics and wine quality: a survey on the performance of 50 commercial active dry yeasts. *Vitis* **2013**, 52, 149−155.

(7) Ferreira, J.; Du Toit, M.; Du Toit, W. J. The effects of copper and high sugar concentrations on growth, fermentation efficiency and volatile acidity production of different commercial wine yeast strains. *Aust. J. Grape Wine Res.* **2006**, 12, 50−56.

(8) Sun, X.; Liu, L.; Zhao, Y.; Ma, T.; Zhao, F.; Huang, W.; Zhan, J. Effect of copper stress on growth characteristics and fermentation properties of *Saccharomyces cerevisiae* and the pathway of copper adsorption during wine fermentation. *Food Chem.* **2016**, 192, 43−52.

(9) Grant, C. M.; MacIver, F. H.; Dawes, I. W. Glutathione is an essential metabolite required for resistance to oxidative stress in the yeast *Saccharomyces cerevisiae*. *Curr. Genet.* **1996**, 29, 511−515.

(10) Penninckx, M. A short review on the role of glutathione in the response of yeasts to nutritional, environmental, and oxidative stresses. *Enzyme Microb. Technol.* **2000**, 26, 737−742.

(11) Miyake, T.; Hazu, T.; Yoshida, S.; Kanayama, M.; Tomochika, K.-i.; Shinoda, S.; Ono, B.-i. Glutathione transport systems of the budding yeast *Saccharomyces cerevisiae*. *Biosci. Biotechnol. Biochem.* 1998, 62, 1858−1864.

(12) Bourbouloux, A.; Shahi, P.; Chakladar, A.; Delrot, S.; Bachhawat, A. K. Hgt1p, a high affinity glutathione transporter from the yeast *Saccharomyces cerevisiae*. *J. Biol. Chem.* **2000**, 275, 13259−13265.

(13) Penninckx, M. J. An overview on glutathione in *Saccharomyces* versus non-conventional yeasts. *FEMS Yeast Res.* **2002**, 2, 295−305.

(14) Grant, C. M.; Dawes, I. W. Synthesis and role of glutathione in protection against oxidative stress in yeast. *Redox Rep.* **1996**, 2, 223−229.

(15) Hedley, D. W.; Chow, S. Evaluation of methods for measuring cellular glutathione content using flow cytometry. *Cytometry* **1994**, 15, 349−358.

(16) Kritzinger, E. C.; Bauer, F. F.; Du Toit, W. J. Role of glutathione in winemaking: a review. *J. Agric. Food Chem.* **2013**, 61, 269−277.

(17) Singleton, V. L.; Salgues, M.; Zaya, J.; Trousdale, E. Caftaric acid disappearance and conversion to products of enzymic oxidation in grape must and wine. *Am. J. Enol. Viticult.* **1985**, 36, 50−56.
(18) Ugliano, M.; Kwiatkowski, M.; Vidal, S.; Capone, D.; Siebert, T.; Dieval, J.-B.; Aagaard, O.; Waters, E. J. Evolution of 3-mercaptohexanol, hydrogen sulfide, and methyl mercaptan during bottle storage of Sauvignon blanc wines. Effect of glutathione, copper, oxygen exposure, and closure-derived oxygen. *J. Agric. Food Chem.* **2011**, 59, 2564−2572.
(19) Sonni, F.; Clark, A. C.; Prenzler, P. D.; Riponi, C.; Scollary, G. R. Antioxidant action of glutathione and the ascorbic acid/glutathione pair in a model white wine. *J. Agric. Food Chem.* **2011**, 59, 3940−3949.
(20) Wegmann-Herr, P.; Berner, M.; Dickescheid, C.; von Waldbrunn, C.; Nickolaus, P.; Durner, D. *Influence of Ascorbic Acid, Sulfur dioxide and glutathione on oxidation product formation in winelike systems*; BIO Web Conferences 5: 38th World Congress of Vine and Wine, 2015.
(21) Paton, A. M.; Jones, S. M. The observation and enumeration of micro-organisms in fluids using membrane filtration and incident fluorescence microscopy. *J. Appl. Bacteriol.* **1975**, 38, 199−200.
(22) Bradford, M. M. A rapid and sensitive method for the quantitation of microgram quantities of protein utilizing the principle of protein-dye binding. *Anal. Biochem.* **1976**, 72, 248−254.
(23) Mauricio, J. C.; Moreno, J. J.; Ortega, J. M. *In vitro* specific activities of alcohol and aldehyde dehydrogenases from two flor yeasts during controlled wine aging. *J. Agric. Food Chem.* **1997**, 45, 1967−1971.
(24) Bisswanger, H. Enzyme assays. In *Practical enzymology*, 2; Wiley-Blackwell: Weinheim, Germany, 2013; pp 93−264.
(25) Zwietering, M. H.; Jongenburger, I.; Rombouts, F. M.; Van't Riet, K. Modeling of the bacterial growth curve. *Appl. Environ. Microbiol.* **1990**, 56, 1875−1881.
(26) Park, S. K.; Boulton, R. B.; Noble, A. C. Automated HPLC analysis of glutathione and thiol-containing compounds in grape juice and wine using pre-column derivatization with fluorescence detection. *Food Chem.* **2000**, 68, 475−480.
(27) Bundesamt für Verbraucherschutz und Lebensmittelsicherheit (BVL). L 00.00-19/2: Untersuchung von Lebensmitteln; Bestimmung von Spurenelementen in Lebensmitteln, Teil 2: Bestimmung von Eisen, Kupfer, Mangan und Zink mit der Atomabsorptionsspektrometrie (AAS) in der Flamme. Beuth Verlag GmbH: Berlin. 2006.
(28) Longin, C.; Petitgonnet, C.; Guilloux-Benatier, M.; Rousseaux, S.; Alexandre, H. Application of flow cytometry to wine microorganisms. *Food Microbiol.* **2017**, 62, 221−231.
(29) Sun, X.; Liu, L.; Ma, T.; Yu, J.; Huang, W.; Fang, Y.; Zhan, J. Effect of high Cu^{2+} stress on fermentation performance and copper biosorption of *Saccharomyces cerevisiae* during wine fermentation. *Food Sci. Technol.* **2019**, 39, 19−26.
(30) Sommer, S.; Wegmann-Herr, P.; Wacker, M.; Fischer, U. Rationale for a stronger disposition of Chardonnay wines for stuck and sluggish fermentation. *S. Afr. J. Enol. Vitic.* **2015**, 36, 180−190.

(31) Sun, X.-y.; Zhao, Y.; Liu, L.-l.; Jia, B.; Zhao, F.; Huang, W.-d.; Zhan, J.-c. Copper tolerance and biosorption of *Saccharomyces cerevisiae* during alcoholic fermentation. *PLoS One* **2015**, 10, No. e0128611.
(32) Vest, K. E.; Zhu, X.; Cobine, P. A. Copper Disposition in Yeast. In *Clinical and Translational Perspectives on WILSON DISEASE*, 1; Kerkar, N., Roberts, E. A., Eds.; Academic Press: London, United Kingdom, 2019; pp 115−126.
(33) Brady, D.; Duncan, J. R. Bioaccumulation of metal cations by *Saccharomyces cerevisiae*. *Appl. Microbiol. Biotechnol.* **1994**, 41, 149−154.
(34) Brandolini, V.; Tedeschi, P.; Capece, A.; Maietti, A.; Mazzotta, D.; Salzano, G.; Paparella, A.; Romano, P. *Saccharomyces cerevisiae* wine strains differing in copper resistance exhibit different capability to reduce copper content in wine. *World J. Microbiol. Biotechnol.* **2002**, 18, 499−503.
(35) Am Ferreira, A. M. D. C.; Ciriolo, M. R.; Marcocci, L.; Rotilio, G. Copper (I) transfer into metallothionein mediated by glutathione. *Biochem. J.* **1993**, 292, 673−676.
(36) Dhaoui, M.; Auchère, F.; Blaiseau, P.-L.; Lesuisse, E.; Landoulsi, A.; Camadro, J.-M.; Haguenauer Tsapis, R.; Belgareh Touzé,N. Gex1 is a yeast glutathione exchanger that interferes with pH and redox homeostasis. *Mol. Biol. Cell.* **2011**, 22, 2054−2067.
(37) Połeć-Pawlak, K.; Ruzik, R.; Lipiec, E. Investigation of Cd (II), Pb (II) and Cu (I) complexation by glutathione and its component amino acids by ESI-MS and size exclusion chromatography coupled to ICP-MS and ESI-MS. *Talanta* **2007**, 72, 1564−1572.
(38) Park, S. K.; Boulton, R. B.; Noble, A. C. Formation of hydrogen sulfide and glutathione during fermentation of white grape musts. *Am. J. Enol. Viticult.* **2000**, 51, 91−97.
(39) Lavigne, V.; Pons, A.; Dubourdieu, D. Assay of glutathione in must and wines using capillary electrophoresis and laser-induced fluorescence detection: changes in concentration in dry white wines during alcoholic fermentation and aging. *J. Chromatogr. A* **2007**, 1139, 130−135.
(40) Du Toit, W. J.; Lisjak, K.; Stander, M.; Prevoo, D. Using LC-MSMS to assess glutathione levels in South African white grape juices and wines made with different levels of oxygen. *J. Agric. Food Chem.* **2007**, 55, 2765−2769.
(41) Maiorella, B.; Blanch, H. W.; Wilke, C. R. By-product inhibition effects on ethanolic fermentation by *Saccharomyces cerevisiae*. *Biotechnol. Bioeng.* **1983**, 25, 103−121.
(42) Walker-Caprioglio, H. M.; Parks, L. W. Autoconditioning factor relieves ethanol-induced growth inhibition of *Saccharomyces cerevisiae*. *Appl. Environ. Microbiol.* **1987**, 53, 33−35.
(43) Stanley, G. A.; Douglas, N. G.; Every, E. J.; Tzanatos, T.; Pamment, N. B. Inhibition and stimulation of yeast growth by acetaldehyde. *Biotechnol. Lett.* **1993**, 15, 1199−1204.
(44) Liu, S.-Q.; Pilone, G. J. An overview of formation and roles of acetaldehyde in winemaking with emphasis on microbiological implications. *Int. J. Food Sci. Technol.* **2000**, 35, 49−61.
(45) Fernández, M. J.; Gómez-Moreno, C.; Ruiz-Amil, M. Induction of isoenzymes of alcohol dehydrogenase in "flor" yeast. *Arch. Microbiol.* **1972**, 84, 153−160.
(46) Wills, C. Production of yeast alcohol dehydrogenase isoenzymes by selection. *Nature* **1976**, 261, 26.

(47) Ciriacy, M. Genetics of alcohol dehydrogenase in *Saccharomyces cerevisiae*: I. Isolation and genetic analysis of adh mutants. *Mutat. Res., Fundam. Mol. Mech. Mutagen*. **1975**, 29, 315−325.
(48) Magonet, E.; Hayen, P.; Delforge, D.; Delaive, E.; Remacle, J. Importance of the structural zinc atom for the stability of yeast alcohol dehydrogenase. *Biochem. J.* **1992**, 287, 361−365.
(49) Cavaletto, M.; Pessione, E.; Vanni, A.; Giunta, C. Improved resistance to transition metals of a cobalt-substituted alcohol dehydrogenase 1 from *Saccharomyces cerevisiae*. J. Biotechnol. 2000, 84, 87−91.
(50) Vanni, A.; Anfossi, L.; Pessione, E.; Giovannoli, C. Catalytic and spectroscopic characterisation of a copper-substituted alcohol dehydrogenase from yeast. *Int. J. Biol. Macromol.* **2002**, 30, 41−45.

Chapter 3

Oxidation of wine polyphenols by secretomes of wild *Botrytis cinerea* strains from white and red grape varieties and determination of their specific laccase activity

Processing of *Botrytis cinerea*-infected grapes leads to enhanced enzymatic browning reactions mainly caused by the enzyme laccase which is able to oxidize a wide range of phenolic compounds. The extent of color deterioration depends on the activity of the enzymes secreted by the fungus. The present study revealed significant differences in the oxidative properties of secretomes of several *B. cinerea* strains isolated from five grape varieties. The presumed laccase-containing secretomes varied in their catalytic activity toward six phenolic compounds present in grapes. All strains led to identical product profiles for five of six substrates, but two strains showed deviating product profiles during gallic acid oxidation. Fast oxidation of caffeic acid, ferulic acid, and malvidin 3-*O*-glucoside was observed. Product formation rates and relative product concentrations were determined. The results reflect the wide range of enzyme activity and the corresponding different impact on color deterioration by *B. cinerea*.

Keywords: *Botrytis cinerea*, laccase, wine polyphenols, oxidation products

---

1 Introduction

Botrytis cinerea infection is a significant problem for the wine industry. As a fungal pathogen, it infects the grapevine (*Vitis vinifera* L.) causing the so-called gray mold.[1,2] The infection process of the grape is accompanied by the secretion of laccase, an enzyme that may cause enzymatic browning of must and wine. After crushing of grapes, oxidation is mainly catalyzed by tyrosinase, which is the grape polyphenoloxidase. Oxidation of hydroxycinnamic acid-tartaric acid esters like caftaric acid results in the formation of *ortho*-quinones that may further condense and polymerize with other phenolic compounds to brown pigments.[3–5] This reaction can be prevented by glutathione (GSH), which traps *o*-quinones of tartaric acid forming the "grape reaction product" (GRP)[3] identified as 2-*S*-glutathionyl caftaric acid.[6] Whereas this product is almost resistant against oxidation by tyrosinase,[3] laccase is able to oxidize a wider range of substrates including GRP.[7,8] Laccase-mediated oxidation of GRP results in the formation of brown polymers or, in the presence of sufficient amounts of GSH, in the formation of 2,5-di-*S*-glutathionyl caftaric acid (GRP2).[9] Besides the presence of laccase and the accompanied oxidation, processing of *Botrytis*-infected grapes entails numerous other problems like high concentrations of glucans, fermentation inhibiting compounds,[10,11] or undesired aroma compounds.[2,12] Whereas some studies demonstrated that the occurrence of laccase might be used as a reliable indicator of *Botrytis* infection,[7,13] others have shown that a high degree of infection is not necessarily accompanied by a high laccase activity.[14,15] Previous studies supposed that the observed differences in enzyme activity can be ascribed to different strains of *B. cinerea*.[16–19] Dubernet et al. obtained a crude preparation of *B. cinerea* laccase, which was able to oxidize a wide range of substrates.[20] Other studies indicated that *B. cinerea* is able to produce different extracellular laccases, depending on the inducing compounds in the culture medium, such as gallic acid or grape juice.[20–24] The produced enzymes exhibit several structural differences in their molecular weight, sugar or amino acid composition, and isoelectric point and show different properties like pH or temperature optimum and substrate specificity.[18,22–24]

The extracellular laccase as a part of the fungal secretome is responsible for the oxidizing activity of *B. cinerea*. The secretomes of 10 *B. cinerea* strains isolated from red and white grape varieties were used to assess the oxidizing properties of the strains. The catalytic activity toward the specific substrate syringaldazine and six common wine phenols was compared between the secretomes and a commercial laccase from *Trametes versicolor*. The product formation during substrate oxidation was observed to characterize the variety of oxidizing properties of *B. cinerea* strains. The results might lead to a better understanding and control of color deterioration caused by laccases, since processing of *B. cinerea*-infected grapes cannot be avoided especially in cold-climate viticulture.

2 Materials and methods

Fungal strains, chemicals, and reagents. Ten *B. cinerea* strains were kindly provided by the Institute of Plant Protection, DLR Rheinpfalz (Neustadt/Weinstrasse, Germany). Five strains had been isolated from red grape varieties (4 Pinot noir, 1 Portugieser) and five from white grape varieties (2 Pinot blanc, 2 Riesling, 1 Chardonnay) during harvest in the Palatinate area (Germany) in 2015 (**Table 3.1**). Commercial *T. versicolor* laccase (*Tra*Lacc) (Sigma-Aldrich, Steinheim, Germany) and *Agaricus bisporus* tyrosinase (PPO) (Worthington Biochemical Corp., Lakewood, NJ) were used for comparison.

Table 3.1: Specific laccase activity in secretomes of *B. cinerea* strains of different grape varieties and of commercial *T. versicolor* laccase. Different letters indicate significant differences[a]

strain	grape variety	specific activity[b] ($U_{Lacc}\cdot mg^{-1}$)
RL01	Riesling	$0.39 \pm 0.14_{b, c}$
RL02	Riesling	$0.14 \pm 0.03_{c, d}$
CD01	Chardonnay	$0.22 \pm 0.04_{b, c, d}$
PB01	Pinot blanc	$0.18 \pm 0.03_{b, c, d}$
PB02	Pinot blanc	$2.33 \pm 0.14_{a}$
PG01	Portugieser	$0.48 \pm 0.07_{b}$
PN01	Pinot noir	$0.23 \pm 0.05_{b, c, d}$
PN02	Pinot noir	$0.40 \pm 0.15_{b, c}$
PN03	Pinot noir	$0.03 \pm 0.01_{d}$
PN04	Pinot noir	$0.05 \pm 0.00_{d}$
*Tra*Lacc	-	$0.13 \pm 0.02_{c, d}$

[a]Different letters indicate significant differences. [b]Determined using the syringaldazine assay and corrected by the total amount of protein in the extract.

For protein preparation, acetone (VWR, Darmstadt, Germany), ethanol absolute >99.9% (Chemsolute, Renningen, Germany), hydrochloric acid ≥32% and Tween 80, both from Carl Roth (Karlsruhe, Germany), were used. Albumin fraction V (from bovine serum), Coomassie Brilliant Blue G-250, and *ortho*-phosphoric acid 85% were obtained from Merck (Darmstadt, Germany). HPLC-grade acetonitrile was from Chemsolute (Renningen, Germany) and formic acid ≥98% from Sigma-Aldrich (Steinheim, Germany). The substrates *p*-coumaric acid >98% and gallic acid >98% were purchased from Fluka (Buchs, Switzerland). Syringaldazine 99%, caffeic acid ≥95%, and (+)-catechin monohydrate ≥98% were from Sigma-Aldrich (Steinheim, Germany). Ferulic acid >98% was from Carl Roth (Karlsruhe, Germany) and malvidin 3-*O*-glucoside chloride >85% from Phytoplan Diehm & Neuberger GmbH (Heidelberg, Germany).

Cultivation of strains. Strains were grown on malt extract agar (30 $g \cdot L^{-1}$) for a minimum of 5 days at 25 °C. Spores were harvested with sterile Tween solution (0.1% v/v) using a Drigalski spatula, and the suspension was filtered through sterile cheesecloth to remove the mycelium. The concentration of spore suspension was assessed using a Fuchs-Rosenthal counting chamber. The spore suspension was stored at -20 °C. To increase protein yield, the following cultivation steps were conducted in triplicate for each *B. cinerea* strain. A volume of 100 mL of medium, containing 20 $g \cdot L^{-1}$ malt extract and 0.1 M citrate phosphate buffer (pH 3.4), was filled in a 500 mL Erlenmeyer flask. Gallic acid (1 $g \cdot L^{-1}$) was added after autoclaving the medium to induce enzyme production. After inoculation of the medium with 10^5 spores, the fungus was cultivated in the presence of the inductor for 3 days under continuous shaking at 25 °C. The fungus was then washed with 0.1 M citrate phosphate buffer (pH 3.4) and centrifuged (7000 *g*, 10 min) to remove the inductor. Cultivation was continued for 11 days in 200 mL of fresh medium.

Preparation of secretomes and determination of protein content. Spores and mycelium were removed by centrifugation (7000 *g*, 10 min). Enzyme precipitation was conducted at -20 °C for 1 h after mixing the supernatant with precooled acetone 2:3 (v/v). The secretome was collected by vacuum filtration. After the solvent was evaporated, the secretome was dissolved in 30 mL of 46 mM sodium acetate buffer (pH 5). The solution was dialyzed with a cutoff of 12–14 kDa against the same buffer overnight at 4 °C. The protein concentration in the secretome preparations was determined using the Bradford assay[25] and bovine serum albumin (BSA) as the protein standard. The protein was dissolved in different concentrations in 0.1 M sodium acetate buffer (pH 5). For calibration, the absorption was measured with a FLUOstar Omega microplate reader spectrophotometer (BMG Labtech, Ortenberg, Germany) at 595 nm and compared with the secretome preparations.

Syringaldazine assay for determination of laccase activity. The determination of the laccase activity in the secretomes was conducted with the laccase-specific substrate syringaldazine according to Grassin and Dubourdieu.[13] The enzyme oxidizes the substrate to the corresponding violet quinone, which is measured photometrically at 530 nm ($\varepsilon_{530\ nm}$ = 65 $mM^{-1} \cdot cm^{-1}$).[26]

A syringaldazine stock solution was prepared with a concentration of 1.5 mM in ethanol, which was diluted with 46 mM sodium acetate buffer (pH 5) to a concentration of 187 µM prior to use. Solutions of secretomes, *Tra*Lacc, and PPO were diluted with buffer at different ratios to a final volume of 100 µL, to which 200 µL of the syringaldazine solution was added to start the reaction. Each sample was prepared in duplicate. One sample without secretome or enzyme and one sample without substrate served as negative controls. The absorption was measured with a FLUOstar Omega spectrophotometer at 25 °C for 1 h in intervals of 30 s and corrected by subtraction of the absorption of a blank consisting of 300 µL of buffer solution. Substrate transformation was calculated by determination of the absorption change per minute. The amount of enzyme oxidizing 1 µmol syringaldazine per minute was designated as 1 unit (U_{Lacc}). Enzyme activity was calculated using the extinction coefficient of syringaldazine. The specific activity was finally obtained by eliminating the dilution factor and by including the protein concentration determined by the Bradford assay. To ensure comparability in the following assays, the volume of secretome solutions and *Tra*Lacc solution were adjusted to yield an amount of 0.001 U_{Lacc}.

UHPLC-DAD analysis of substrate oxidation. UHPLC-DAD analysis was used to examine the decrease in substrate concentration and the formation of products as a consequence of enzymatic oxidation and consecutive reactions. Stock solutions of each substrate were prepared and all substrates were applied in a final concentration of

0.2 mM, except gallic acid and (+)-catechin, which were oxidized in concentrations of 1.5 mM. Aliquots of 1 mL of substrate solution were filled into reaction vials and 0.5 mL of secretome or *Tra*Lacc solution (0.001 U_{Lacc}) was added. In case of PPO solution, 0.1 U_{PPO} was applied. Because the catalytic activity of the PPO was assessed only qualitatively, 0.1 U_{PPO} was chosen on the basis of preliminary tests. This amount led to a sufficient substrate oxidation within 48 h. The vials were made up to a total volume of 5 mL with 46 M sodium acetate buffer (pH 5), or 0.1 M citrate phosphate buffer (pH 3.4) in the case of malvidin 3-*O*-glucoside. A negative control was prepared with no added secretome or enzyme solution. Reaction mixtures were incubated at 37 °C for 48 h, and samples of 700 µL were taken at 0, 0.25, 1, 4, 24, and 48 h. The reaction was stopped by the addition of 175 µL 3% (v/v) hydrochloric acid. All experiments were carried out in duplicate.

Samples were analyzed with a Prominence UFLC system (Shimadzu, Kyoto, Japan) equipped with a Prominence DGU-20A5R degasser, two Nexera X2 LC-30AD pumps, a Nexera SIL-30AC Prominence autosampler, an Acquity UPLC HSS T3-column (150 x 2.1 mm, particle size 1.8 µm) from Waters (Milford, MA), a CTO-20AC Prominence column oven, and a SPD-M20A Prominence DAD detector. The analysis of substrate oxidation of ferulic acid, caffeic acid, (+)-catechin, and gallic acid was conducted with the following gradient program at a flow rate of 0.4 mL·min^{-1} using solvent A 0.1% (v/v) formic acid in water and solvent B 0.1% (v/v) formic acid in acetonitrile: 0 min, 2% B; 20 min, 40% B; 21 min, 100% B; 25 min, 100% B; 26 min, 2% B; 30 min, 2% B. For the analysis of samples containing malvidin 3-*O*-glucoside and its oxidation products, the solvent composition was changed to solvent A 5% (v/v) formic acid in water and solvent B 5% (v/v) formic acid in acetonitrile, and the following gradient program was applied: 0 min, 4% B; 2 min, 4% B; 7 min, 8% B; 13 min, 10% B; 19 min, 17% B; 23 min, 30% B; 23.3 min, 100% B; 25.3 min, 100% B; 25.8 min, 4% B. The column temperature was 40 °C, and the injection volume was 5 µL. Substrates were quantified using commercial standards.

UHPLC-ESI-MS^n identification of oxidation products. The identification of products was conducted on a Waters Acquity I-Class system (Milford, MA) coupled with an LTQ-XL ion trap mass spectrometer (Thermo Scientific, Inc., Braunschweig, Germany). The

Table 3.2: Substrate half-life ($t_{1/2}$) for oxidation with secretomes of *B. cinerea* strains and *T. versicolor* laccase. Different letters indicate significant differences.

	(+)-catechin	ferulic acid	caffeic acid	gallic acid	*p*-coumaric acid	malvidin 3-*O*-glucoside
strain	$t_{1/2}$ (h)	$t_{1/2}$ (h)	$t_{1/2}$ (h)	$t_{1/2}$ (h)	$t_{1/2}$ (h)	$t_{1/2}$ (h)
RL01	$5.1 \pm 1.7_{c,\,d}$	$2.1 \pm 0.5_{b,\,c}$	$0.5 \pm 0.0_{a}$	$7.4 \pm 0.4_{d}$	$6.5 \pm 0.9_{b,\,c}$	$1.9 \pm 0.1_{b,\,c}$
RL02	$10.9 \pm 0.8_{b}$	$0.9 \pm 0.1_{d,\,e}$	$1.5 \pm 0.2_{a}$	$14.2 \pm 0.3_{b}$	$6.0 \pm 0.9_{b,\,c}$	$1.6 \pm 0.1_{b,\,c}$
CD01	$6.9 \pm 0.2_{b,\,c,\,d}$	$2.7 \pm 0.3_{b}$	$2.1 \pm 0.1_{a}$	$10.8 \pm 0.7_{c}$	$7.9 \pm 0.1_{b}$	$3.7 \pm 0.2_{a}$
PB01	$10.8 \pm 2.8_{b}$	$2.8 \pm 0.4_{b}$	$1.9 \pm 1.4_{a}$	$22.2 \pm 1.8_{a}$	$11.5 \pm 1.8_{a}$	$1.6 \pm 0.1_{b,\,c}$
PB02	$16.1 \pm 0.0_{a}$	$4.0 \pm 0.1_{a}$	$1.8 \pm 0.0_{a}$	$21.0 \pm 0.4_{a}$	$8.3 \pm 0.0_{b}$	$2.3 \pm 0.0_{a,\,b,\,c}$
PG01	$9.7 \pm 1.1_{b,\,c}$	$1.7 \pm 0.2_{c,\,d}$	$1.1 \pm 0.0_{a}$	$11.6 \pm 0.3_{b,\,c}$	$9.1 \pm 0.4_{a,\,b}$	$2.5 \pm 0.2_{a,\,b,\,c}$
PN01	$6.6 \pm 0.0_{b,\,c,\,d}$	$0.5 \pm 0.0_{e}$	$0.4 \pm 0.0_{a}$	$11.0 \pm 0.6_{b,\,c}$	$6.8 \pm 0.7_{b,\,c}$	$1.0 \pm 0.2_{c}$
PN02	$8.3 \pm 0.4_{b,\,c}$	$0.9 \pm 0.0_{d,\,e}$	$1.1 \pm 0.0_{a}$	$12.2 \pm 0.1_{b,\,c}$	$4.7 \pm 0.3_{c,\,d}$	$2.2 \pm 0.3_{a,\,b,\,c}$
*Tra*Lacc	$3.1 \pm 0.2_{d}$	$0.1 \pm 0.0_{e}$	$0.3 \pm 0.0_{a}$	$4.8 \pm 0.6_{d}$	$2.5 \pm 0.1_{d}$	$3.1 \pm 1.1_{a,\,b}$

same column, eluents, and gradients were applied as described for UHPLC-DAD analysis. The following conditions of the mass spectrometer were used for the analysis of oxidation products except for those derived from malvidin 3-*O*-glucoside: the capillary temperature was set at 300 °C in the negative electrospray ionization (ESI) mode and run at a voltage of 28 V. The source voltage was kept at 4 kV at a current of 100 µA. The tube lens was set at 125 V. Nitrogen was used as the sheath, auxiliary, and sweep gas at a flow of 50, 5, and 1 arbitrary units, respectively. Malvidin 3-*O*-glucoside products were identified in the positive ionization mode with the following conditions: the temperature of the capillary was set at 325 °C at a voltage of -25 V. The source voltage was held at 3 kV at a current of 100 µA. The tube lens was adjusted to -65 V. Nitrogen gas flow differed with 60, 8, and 1 arbitrary units. In both methods, collision-induced dissociation spectra were received at 35 eV using helium as the collision gas.

Evaluation of substrate oxidation. Oxidation by *B. cinerea* secretomes and *Tra*Lacc was assessed on the basis of substrate half-life ($t_{1/2}$). The data obtained was normalized and fitted to a first-order kinetic model using the software OriginPro 8G (OriginLab Corporation, Northampton, MA). Calculated reaction constants were used to determine substrate half-life. For each substrate, the relative concentration of products was determined by referring it to the product with the highest peak area, which was set at 100%.

Statistical analysis. Statistical analysis of the results was conducted with XLSTAT (Version 2014.4.06, AddinSoft Technologies, Paris, France). In the case of variance homogeneity, an ANOVA (analysis of variance) was used followed by a Tukey-test with a selected significance level of $p < 0.5$.

3 Results and discussion

Comparison of specific laccase activities in *B. cinerea* secretomes. The specific activity of laccase containing secretomes secreted by *B. cinerea* strains was assessed using the syringaldazine assay. The calculated specific activities were corrected by the total amount of protein in the extract.

All extracts showed remarkably different activities (**Table 3.1**) with strain PB02 having the highest activity (2.33 $U_{Lacc}{\cdot}mg^{-1}$). Lowest activities were found for the strains PN03 and PN04. Vortkamp et al. described activity levels of 12 and 1.44 $U_{Lacc}{\cdot}mg^{-1}$ for two induced *B. cinerea* laccases.[19] Grassin and Dubourdieu observed maximum levels of 0.6 $U_{Lacc}{\cdot}mL^{-1}$.[13] However, analyses in these studies differed from the present study since laccase production was induced for up to 25 days19 and activity was measured without previous enzyme precipitation and dialysis.[13,19]

Gallic acid was described as the most effective inducer of laccase secretion compared to other phenolic compounds.[21] Maximum levels of extracellular laccase in the medium were obtained when gallic acid was continuously present.[23] In order to analyze enzyme activity toward syringaldazine, gallic acid needed to be removed in the present study because it would have interfered with the assay. An induction period of 3 days was chosen, since important processes of enzyme production have been shown to occur during the first 2–3 days of fungal growth when the inductor is available. Later addition of the inductor did not increase laccase activity.[21,23] The activity of induced laccase obviously depends on the individual *B. cinerea* strain, due to genetic variability. Several studies already confirmed variability between strains, resulting in differences in growth rates, sclerotium production and pathogenicity, or in altered production of enzymes *in vitro* and *in vivo*.[27–31] The extremely low specific activities of two *B. cinerea* strains from Pinot noir grapes might be attributed to mutations, resulting in the production of mostly inactive forms of laccases. Previous studies demonstrated that mutations introduced at different sites of fungal laccases led to a reduced catalytic activity of the enzyme or changes in pH optima and substrate affinities.[32–34] Furthermore, different *B. cinerea* strains might respond differently to the inductor gallic acid. It has been shown that a product of *in vivo* gallic acid oxidation by *B. cinerea*, secreted to the external medium, actually causes the laccase inducing mechanism.[35] Because fungi are also able to produce a constitutive form of laccase,[36,37] it was assumed that this form may be responsible for gallic acid conversion in the mycelium.[35] The ability of the *B. cinerea* strains' constitutive laccase to metabolize gallic acid might therefore be the reason for the different activities of extracellular laccases in the secretomes. The tested *Tra*Lacc showed a low specific activity, although it has a high purity according to the manufacturer. As expected, no activity of the PPO was measured toward the laccase-specific substrate.

Analysis of substrate oxidation and product formation. Substrate oxidation by different *B. cinerea* strains might be ascribed to laccase activity in the secretomes because the laccase-specific substrate syringaldazine was oxidized applying secretomes. It is likely that other enzymes are present in the secretomes of *B. cinerea* because no purification of laccase was conducted. The fungus secretes various enzymes with functions of degrading plant cell walls for penetration, further host damage, and nutrient consumption. Several of these enzymes were described to be involved in winemaking.[38] The activity of the majority of those enzymes can be excluded in the present study because the subsequent substrate oxidation experiments were conducted in buffer solutions which did not contain suitable substrates for other enzymes but laccase. One of the enzymes that might be found in the secretomes is glucose oxidase, which was recently found to be responsible for an indirect oxidation of polyphenols in wine

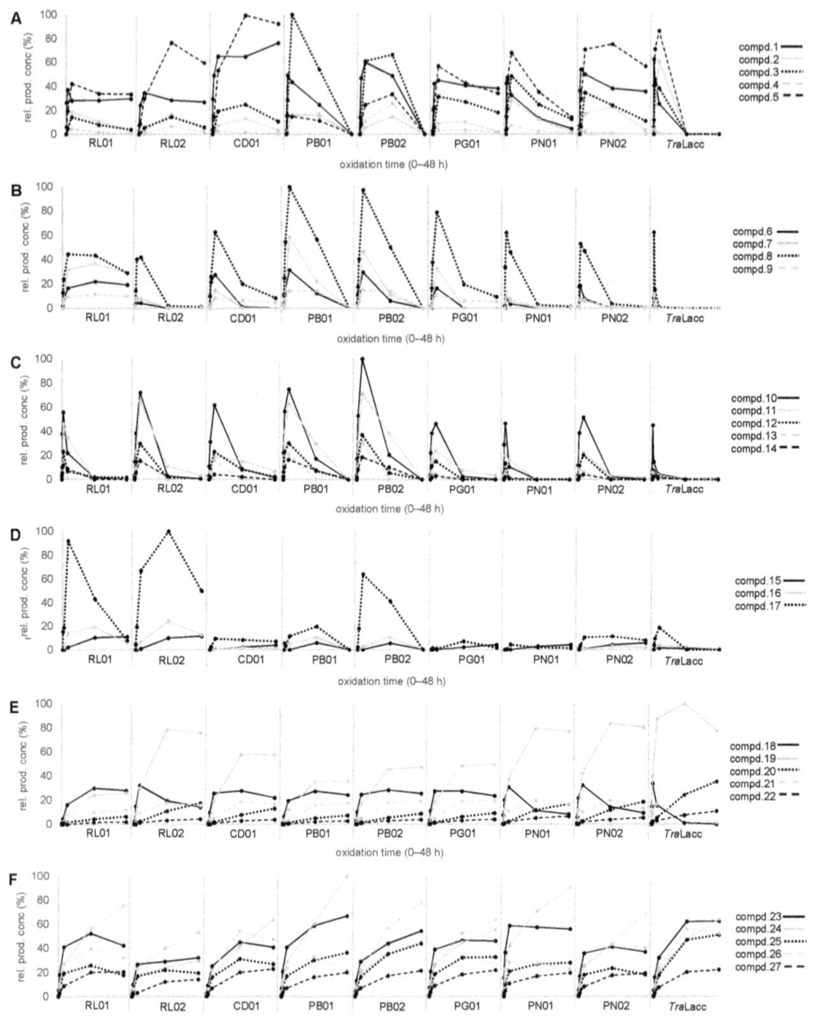

Figure 3.1 Product formation during oxidation of different wine phenols. For each secretome of *B. cinerea* strains and commercial *T. versicolor* laccase, relative product concentrations (referred to the product with the highest peak area) are shown after 0, 0.25, 1, 4, 24, and 48 h of oxidation.; (A) (+)-catechin, (B) ferulic acid, (C) caffeic acid, (D) gallic acid, (E) *p*-coumaric acid, (F) malvidin 3-*O*-glucoside; average standard deviation (n = 2) is 2.9%.

induced by the formation of H_2O_2.[39] Because the oxidation buffers did not contain glucose, this indirect oxidation can be ruled out. The use of laccase as an indicator for *Botrytis* infection and its broad substrate specificity[7,13] indicate that it is the main oxidizing enzyme

Table 3.3: UV-Vis spectra and MS data of products formed during apparent laccase-catalyzed oxidations.

compd. no.	identification	λ_{max}	precursor ion (*m/z*)	fragment ions (*m/z*)	reference
(+)-catechin[a]					
1	catechin dimer	281	577	393, 559, 425, 439, 533, 269	41,42
2	catechin dimer	415	575	n.d.	41
3	catechin dimer	415	575	449, 437, 531, 557, 287, 407	41
4	unknown	320	n.d.		
5	catechin dimer	390	575	449, 437, 394, 287	41,43,44
ferulic acid[a]					
6	ferulic acid dimer	334	385	282, 217, 326, 203, 159, 173	45
7	ferulic acid dimer (−CO_2)	334	341	297, 282, 173	45
8	ferulic acid dimer	328	385	341, 249	45
9	ferulic acid trimer (−CO)	290	551	193, 178	
caffeic acid[a]					
10	caffeic acid dimer (−H_2O, −CO)	328	313	269, 295	
11	caffeic acid dimer (−H_2O, −CO)	318	313	269, 179, 135	
12	caffeic acid dimer (−H_2O, −CO)	317	313	269, 179, 135	
13	caffeic acid dimer	334	357	313	
14	caffeic acid dimer	283	357	159, 269, 109	
gallic acid[a]					
15	unknown	262	389	345, 371, 317	
16	unknown	380	447	429, 385	
17	unknown	274	429	385, 341	
p-coumaric acid[a]					
18	*p*-coumaric acid dimer	305	325	281, 237	
19	*p*-coumaric acid dimer	316	325	281, 237	
20	unknown	284	n.d.		
21	unknown	320	n.d.		
22	unknown	296	n.d.		
malvidin 3-*O*-glucoside[b]					
23	unknown	334	379[c]	217, 185, 351	
24	2,4,6-trihydroxybenzaldehyde	292	155	127, 155, 109, 137	55
25	anthocyanone A	292	367[c]	185, 205, 187, 307	55
26	syringic acid	274	199	155, 140, 123	55
27	vanillic acid	290	169	n.d.	

[a]detection in the negative ionization mode, [b]detection in the positive ionization mode, [c]$[M+Na]^+$

secreted by the fungus. It has been assumed that it protects the fungus during the infection process against defense mechanisms of the host like phenolic compounds and phytoalexins.[24] The same mechanism was exploited by the addition of gallic acid as inducing agent to ensure the production and presence of laccase in the secretomes.

Strains PN03 and PN04 showed a very low specific activity, and, due to the limited amount of available protein extract, it was not possible to apply the corresponding secretomes in an amount of 0.001 U_{Lacc}. Both secretomes were not considered in the analysis of substrate oxidation. PPO was only assessed regarding its general ability to oxidize the substrates because the missing activity toward syringaldazine did not permit the application of this enzyme in a comparable amount.

Considerable differences between the catalytic activities of the other *B. cinerea* strains toward phenolic substrates were revealed by determination of the substrate half-life (**Table 3.2**). This can be associated with differences in substrate specificity already reported for induced forms of *B. cinerea* laccases by several authors.[20,23,24] A high specificity leads to a faster substrate conversion to the corresponding quinones. The high specific activity of strain PB02 toward syringaldazine was not accompanied by a faster oxidation of the other substrates compared to the other *B. cinerea* strains.

Previous studies already attributed differences in substrate specificity to laccases of different *B. cinerea* strains isolated from grapes.[17,19] Furthermore, different pH and temperature optima have been observed.[17] The use of constant reaction conditions that might differ from the optimum condition of the enzymes might affect the catalytic activity of different *B. cinerea* secretomes. The enzymatic oxidation of phenols usually takes place at low pH values because of the natural occurrence of *B. cinerea* on fruits.[36,37] The limited solubility of substrates, however, did not allow the use of pH values lower than 5 in the present study except for the oxidation of malvidin 3-*O*-glucoside.

Oxidation by different *B. cinerea* secretomes led to considerable differences in the change of relative product concentrations over time (**Figure 3.1 A–F**). Whereas the rate of product formation depends mainly on the catalytic activity of the enzymes, the decrease in relative product concentrations is a result of subsequent, non-enzymatic reactions of the highly reactive quinones. Therefore, the latter reaction can be considered as an enzyme-independent reaction. The resulting formation of substrate oligomers and further polymerization led to considerable changes in the color of the reaction mixtures. Solutions turned to a yellow, greenish-yellow, or brown color during oxidation of (+)-catechin, gallic acid, and caffeic acid, respectively. In contrast, oxidation of malvidin 3-*O*-glucoside resulted in a discoloration. These reaction mechanisms may also occur in wine and thus adversely affect wine color. Maximum concentrations of the products are listed in **Table S 3.1** in the supporting information.

Oxidation of (+)-catechin. The UHPLC analysis showed five additional peaks formed by the oxidation of catechin (compounds **1–5**, **Table 3.3**). Molecular ions with *m/z* 575 and 577 indicate the presence of catechin dimers. These were previously described for (+)-catechin oxidation by a crude grape polyphenoloxidase.[40,41] A previous study reported seven major peaks formed at low pH values (pH 3) and three major peaks formed at high pH values (pH 6).[40] Two types of dimers have been described for the enzymatic oxidation of (+)-catechin, colorless dehydrocatechins B[42] and yellow colored dehydrocatechins A.[43,44] Dehydrocatechins A are predominantly formed at higher pH.[41] Because the present tests were conducted at pH 5, formation of yellow products is likely and is manifested in the development of a clear yellow color of the solution. Yellow catechin dimers, identified after reaction with grape PPO, showed distinct absorption maxima at 256, 280, 385, and 412 nm, whereas colorless products showed the same absorption spectrum as (+)-catechin with a maximum of 280 nm.[40] Compound **1** was identified as dehydrocatechin B based on data of previous studies.[40,41] This compound eluted prior to (+)-catechin with an absorption maximum at 281 nm and fragment ions at *m/z* 439 and 425. Absorption maxima of compounds **2**, **3**, and **5** correspond to the yellow derivatives. The fragment ion at *m/z* 394 has already been detected in a previous study and was described as the product of the loss of the benzopyran moiety of the upper catechin unit.[41] Compound **5** can therefore be tentatively identified as dehydrocatechin A.

Substrate half-life of (+)-catechin ranged from 3.1–16.1 h. *Tra*Lacc catalyzed oxidation showed the lowest $t_{1/2}$ value. Among the strains, oxidation with secretomes of RL01, CD01, and PN01 caused the shortest substrate half-life. The rate of product formation differed between strains (**Figure 3.1 A**). Oxidation with RL01, PB01, PG01, PN01, and PN02 showed a fast product formation, reaching the maximum in relative product concentrations after 4 h. Apparent laccase activity of the strains RL02, CD01, and PB02 generally led to a slower product formation with maximum concentrations not before 24 h. The fastest product formation was observed for the oxidation with *Tra*Lacc, showing the highest concentrations already after 0.25 to 4 h and products were completely depleted after 24 h. Oxidation with both Pinot blanc isolates differed from those of other *B. cinerea* strains. As observed for *Tra*Lacc, oxidation products were highly instable and completely diminished after 48 h. The predominant product of most secretomes was compound **5**, whereas PB01 and PB02 led to higher concentrations of compound **3**.

Oxidation of ferulic acid. The UHPLC analysis revealed four products with absorption spectra similar to ferulic acid (compounds **6–9**, **Table 3.3**). Two products have been identified after oxidation of ferulic acid with a laccase from *Pyricularia oryzae* and described as ferulic acid dimers with molecular ion peaks at *m/z* 385 and fragment ions at *m/z* 341.[45] The same *m/z* ratios were observed for compounds **6–8**. Reaction mixtures did not show any color change during oxidation.

Substrate half-life values indicate that ferulic acid is readily oxidized by all strains, suggesting that it is a good substrate of all laccases. The lowest half-life values were observed for oxidation with *Tra*Lacc and strains RL02, PN01, and PN02 with $t_{1/2}$ values below 1 h. It has already been demonstrated that laccases of different basidiomycetes exhibit high affinities toward ferulic acid,[46] which can be ascribed to the structure of the substrate. A higher substrate affinity and reaction rate for methoxy-substituted monophenols in comparison to diphenols without substituents has previously been described.[47] A positive inductive effect of the methoxy group as *o*-substituent can stabilize the formed radical by delocalization of electrons and thus increase the catalytic effectiveness.[46,48] This results in a faster conversion of ferulic acid compared to a monophenol like *p*-coumaric acid. This is reflected by the change in relative product concentrations over time (**Figure 3.1 B**). The fast increase is followed by a fast decrease in the concentrations of all products. Only strain RL01 led to a slower product formation and subsequent decrease in the relative concentrations. The predominant product of all catalyzed oxidations was compound **8**. Oxidation with strains RL02, PN01, and PN02, which led to the lowest substrate half-life, showed a faster maximum product formation within 1 h compared to other *B. cinerea* strains, which showed product maxima within 4 h.

The faster product formation is accompanied by lower maximum product concentrations and an earlier decrease of products. The slow product formation of the secretome of RL01 showed product maxima after 24 h. The fastest formation rate for all products was observed during oxidation with *Tra*Lacc. Due to the fast substrate oxidation ($t_{1/2}$ = 0.1 h), the maxima in relative product concentrations were already detected after 0.25 h. This was followed by a total depletion of products after 4 h.

The UHPLC analysis confirmed that ferulic acid does not represent a substrate for the PPO because no substrate decrease and product formation was observed. Ferulic acid has been described as a PPO inhibitor, possibly as a result of interaction with the active site of the enzyme.[8,49–51] Substituents of the phenolic ring and their position can affect the extent of the inhibitory properties of substrates toward PPO.[49]

Oxidation of caffeic acid. UHPLC chromatograms showed a product profile of five caffeic acid dimers (compounds **10–14**) with absorption spectra similar to that of caffeic acid (**Table 3.3**). Božič et al. observed a new absorption band between 450–550 nm in a laccase-catalyzed oxidation of caffeic acid applying spectrometry.[52] In the present study, no product peaks were detected, which have an absorption maximum in this range of wavelengths. This might be explained by the lower substrate concentration. Solutions showed a brown color. Brown has been described as a characteristic color of polymers resulting from caffeic acid oxidation.[4]

A short substrate half-life without significant differences was observed for all oxidations. Compared to other tested substrates, caffeic acid oxidation proceeded fastest. Previous studies have already described a high substrate specificity of different fungal laccases for caffeic acid.[20,24,36,47] This can also be seen in a fast increase in relative product concentrations that is followed by fast product decrease (**Figure 3.1 C**). The most abundant product was compound **10**, which was formed in highest concentrations during all oxidations. Differences can be seen for oxidation by RL01 and PN01, which led to a comparably low substrate half-life below 1 h. Both showed a maximum increase in relative product concentrations within 1 h and products almost completely decreased after 24 h. Other *B. cinerea* secretomes required 4 h for the formation of maximum product concentrations and the following decrease was slower. The highest catalytic activity of

*Tra*Lacc was again reflected in the fastest product formation within 0.25–1 h with a total depletion of products after 24 h.

Oxidation of gallic acid. Gallic acid oxidation represents an exception among the tested substrates because different product profiles were observed in UHPLC chromatograms applying different *B. cinerea* secretomes. Most strains led to three products (compounds **15–17**, **Table 3.3**), but compound **16** could not be detected for PN01 and PG01. Compound **17** was formed in low concentrations via autoxidation of gallic acid in the negative control and might be a gallic acid dimer.[53] Because the concentrations were very low, identification of products was not possible. Absorption maximum of gallic acid oligomers and polymers has been described to be in the range of 390–600 nm after reaction with a *T. versicolor* laccase.[52] Because of the high degree of polymerization and the associated complexity of the polymers, no distinct product peaks with an absorption maximum in this range of wavelengths could be detected. A greenish-brown color of reaction mixtures suggested the presence of polymers.

All strains led to the longest substrate half-life for gallic acid compared to the other substrates. Thus, the third hydroxyl group might decelerate oxidation. A maximum value of 22.2 h was observed for the oxidation with strain PB01. Relatively low $t_{1/2}$ values of 4.8 and 7.4 h were only found for the oxidation with *Tra*Lacc and RL01, respectively. Although enzyme production of *B. cinerea* strains was induced with gallic acid, it does not seem to be a specific substrate for laccases in the secretomes. This can also be deduced from product formation (**Figure 3.1 D**) with generally low maxima in relative product concentrations except compound **17**, which was formed in considerably higher concentrations during oxidation by the strains RL01, RL02, and PB02. A low stability of products might impede the determination of product concentrations since even oxidation with *Tra*Lacc, which led to the shortest substrate half-life, showed low relative product concentrations.

Previous studies reported a high substrate specificity of gallic acid-induced *B. cinerea* laccases for gallic acid.[20,23,24] Another study demonstrated that the specificity for gallic and ferulic acids of grape juice-induced *B. cinerea* laccase did not differ.[20] In contrast, the present study revealed a higher apparent activity of laccases toward ferulic acid reflected in lower $t_{1/2}$ values than for gallic acid. The results for *Tra*Lacc correspond to those from a previous study that analyzed two forms of *T. versicolor* laccases.[36] A lower activity was determined toward gallic acid in comparison to ferulic and caffeic acids, which was confirmed by $t_{1/2}$ values in the present study.

Oxidation of *p*-coumaric acid. UHPLC chromatograms showed five product peaks (compounds **18–22**, **Table 3.3**). Compounds **18** and **19** were identified as *p*-coumaric acid dimers. Due to the low peak areas, compounds **20–22** could not be identified. Oxidation did not lead to a change in color of the reaction mixtures.

*Tra*Lacc ($t_{1/2}$ = 2.5 h) and PN02 ($t_{1/2}$ = 4.7 h) required the shortest time to oxidize half of the initial substrate concentration. Other $t_{1/2}$ values ranged from 6.0–11.5 h. As mentioned above, monophenols were described as inappropriate substrates for laccases in contrast to diphenols and methoxy-substituted monophenols.[23,47] It is possible that *p*-coumaric acid is transformed into caffeic acid by hydroxylation and subsequently oxidized to the corresponding quinone. A previous study demonstrated the lowest relative activity of a *B. cinerea* laccase crude extract toward *p*-coumaric acid, regarding only the substrates also employed in the present study.[20] These results could not completely be verified in the present study because apparent laccase activity toward *p*-coumaric acid and (+)-catechin was comparable. Gallic acid showed even longer substrate half-life values. A study with a laccase from *Lentinus edodes* has demonstrated the lowest substrate affinity toward *p*-coumaric acid in comparison with ferulic and caffeic acids.[47] This corresponds to the substrate half-life in present study, showing a decrease in the catalytic activity toward the substrates in the same order.

The rate of product formation (**Figure 3.1 E**) also reflects the lower apparent laccase activity toward *p*-coumaric acid in comparison to ferulic and caffeic acids. The predominant compound **19** was formed slowly, showing the highest relative product concentrations not before 24 to 48 h. In contrast to the majority of strains, oxidation by RL01 led to a lower concentration of compound **19**, and compound **18** was the most abundant compound formed. Further differences between *B. cinerea* strains were observed. Oxidation with RL02, PN01, and PN02, showing substrate half-life values in a lower range, led to a maximum product increase in compound **18** until 4 h and compound **19** until 24 h, with distinctively higher maximum concentrations of compound **19**. Other *B. cinerea* strains caused a maximum product formation after 24 h for compound **18** and after 48 h for compound **19**. The fastest increase in both compounds was observed for the oxidation with *Tra*Lacc, according to the lowest substrate half-life determined for this oxidation.

Oxidation of malvidin 3-*O*-glucoside. Malvidin 3-*O*-glucoside oxidation led to five product peaks in the UHPLC chromatograms (compounds **23–27**, **Table 3.3**). All *B. cinerea* strains oxidized this *o*-methylated anthocyanin, resulting in a discoloration of the solutions, whereas PPO did not lead to oxidation of the substrate. The B-ring substitution pattern of malvidin is identical to the pattern of syringaldazine, which was also not oxidized by PPO, showing that *o*-methylated compounds are no substrates for PPO. Also steric hindrance of oxidation by the sugar moiety might contribute to the PPO resistance of malvidin, since aglycones are often oxidized by PPO.[54] Traces of compounds **25** to **27** were detected after oxidation with PPO, which might also be attributed to autoxidation of the substrate. Most of the detected compounds are similar to the products of malvidin 3-*O*-glucoside autoxidation identified in O_2 rich model wine solutions.[55] Compounds **24** and **26** were identified as 2,4,6-trihydroxybenzaldehyde and syringic acid, respectively. The latter can be considered a specific degradation product of malvidin, resulting from the B-ring after oxidation cleavage. The substrate employed in this study contained small amounts of peonidin 3-*O*-glucoside. Compound **27** was identified as vanillic acid, which is the corresponding B-ring moiety of this anthocyanin. Based on data reported by Lopes et al., compound **25** was identified as anthocyanone A,[55] which is a nonspecific degradation product since it might be formed by oxidation of both anthocyanins. Compound **23** could not be identified but the similarity in fragmentation pattern suggests a similar structure like anthocyanone A.

Although anthocyanins are only present in red grapes, relatively low substrate half-life values were determined for *B. cinerea* strains isolated from white grape varieties, but the lowest $t_{1/2}$ value of 1.0 h was observed for the oxidation by PN01, a *B. cinerea* strain from Pinot noir grapes. Other values ranged from 1.6−3.7 h. In contrast to the oxidation of the other substrates, *Tra*Lacc caused the highest substrate half-life value for malvidin 3-*O*-glucoside. This might be explained by the lower pH value used for this substrate. It seems likely that *B. cinerea* enzymes like laccase are adapted to the acidity of grapes, whereas *Tra*Lacc, as an enzyme of a white-rot fungus that is growing on wood and litter, might have a higher pH optimum. It has previously been shown that several fungal laccases from basidiomycetes, ascomycetes, and deuteromycetes have different pH optima for the oxidation of different substrates.[37] *Tra*Lacc might have a considerably higher pH optimum for the oxidation of malvidin 3-*O*-gluoside compared to the oxidation of the other substrates.

The comparable range of substrate half-life for malvidin 3-*O*-glucoside and caffeic acid observed in this study is in accordance with a previous study with *B. cinerea* laccases that showed a similar apparent specificity for both substrates.[19] However, the change of relative product concentrations over time for malvidin 3-*O*-glucoside oxidation (**Figure 3.1 F**) differed from that observed for caffeic acid oxidation. Maximum product concentrations were formed slowly by all strains and the products seem to be more stable. All products were found in relatively high concentrations and increased until 24–48 h of oxidation. Unlike products of other substrate oxidations that decreased due to generally fast subsequent chemical reactions, vanillic and syringic acids also represent substrates of *B. cinerea* and *T. versicolor* laccases.[20,36,37,56] In the present study, the missing or slow decrease in concentrations of syringic and vanillic acids indicates an apparent low catalytic activity of laccases toward these substrates.

The compound formed in the highest concentration differed between strains. Syringic acid was found in highest concentrations during oxidation with RL02, PB02, and PG01. All other strains led to highest concentrations of trihydroxybenzaldehyde. Product formation by oxidation with PN01 differed from those of other strains (**Figure 3.1 F**). This strain showed the lowest substrate half-life and led to a faster formation of compound **23** and syringic acid with maximum concentrations already reached after 4 h. Other strains showed maximum product concentrations after 48 h except RL01, CD01, and PN02 that led to maximum concentrations of compounds **23**, **25**, and **26** already after 24 h.

Because inductive laccases are produced and secreted by *B. cinerea* to oxidize phenolic compounds present in the grape, a diversity in substrate specificity of laccases might be a consequence of the polyphenol profile of the grape on which the fungus was found.

Differences in substrate specificity, as a result of different inducing agents for laccase production by *B. cinerea*, have already been characterized in previous studies.[23,24] The presumed laccase-containing secretomes of strains found on the same grape variety did not show similar levels of specific activity or a comparable catalytic activity toward the same substrates. Since product profiles vary between the strains, the influence of the latter on the formation reaction of the distinct products still needs to be evaluated. The understanding of differences in oxidizing activity of *B. cinerea* laccases as a part of the fungal secretome and an elucidation of influencing parameters may support a better control of *B. cinerea* infection of grapes. This provides a basis for limiting deterioration capacity of laccases of wine aroma and color.

Funding

This research project was financially supported by the German Ministry of Economics and Technology (via AiF) and the FEI (Forschungskreis der Ernährungsindustrie e.V., Bonn). Project AiF 18645N.

Acknowledgments

We thank Dr. Andreas Kortekamp of the Institute of Plant Protection (DLR Rheinpfalz, Neustadt/Weinstrasse, Germany) for providing isolated strains of *B. cinerea* and further support.

Supporting information

Table S 3.1: Maximum concentrations of products formed during apparent laccase-catalyzed oxidations.

compd. no.	identification	product concentration (μM)								
(+)-catechin equivalent		RL01	RL02	CD01	PB01	PB02	PG01	PN01	PN02	*Tra*Lacc
1	catechin dimer	19.8±1.3	18.6±0.3	41.0±6.1	26.4±6.3	32.3±1.6	24.1±2.8	25.6±2.4	29.1±1.0	33.8±0.5
2	catechin dimer	5.6±0.2	8.9±1.9	6.9±2.7	8.5±1.4	7.9±0.2	5.5±0.3	15.0±4.6	12.1±1.5	32.9±11.6
3	catechin dimer	4.1±0.7	7.7±2.0	13.3±3.9	53.8±14.0	35.8±2.1	16.8±2.7	26.2±6.9	18.6±6.0	23.2±4.4
4	unknown	2.3±0.4	3.5±2.4	1.3±0.3	9.2±0.0	13.7±2.8	1.8±0.2	3.9±2.9	1.6±0.4	1.7±0.4
5	catechin dimer	22.6±0.3	41.2±31.4	53.5±46.5	8.3±1.0	17.8±4.9	30.6±24.7	36.8±28.0	40.5±36.5	46.4±41.7
ferulic acid equivalent										
6	ferulic acid dimer	4.5±0.8	1.0±0.0	5.6±0.2	6.4±0.2	6.1±0.3	3.4±0.1	1.7±0.0	3.8±0.1	1.8±0.2
7	ferulic acid dimer	7.5±0.5	1.9±0.1	3.0±0.1	11.9±1.4	9.5±0.3	6.7±0.3	2.8±0.1	1.2±0.2	0.6±0.0
8	ferulic acid dimer	8.8±2.4	8.5±2.3	12.7±0.7	20.3±1.8	19.7±2.4	16.0±1.9	12.6±0.6	10.7±1.7	12.6±1.3
9	ferulic acid trimer	2.4±0.6	1.6±0.6	1.7±0.4	3.0±0.3	3.1±0.6	2.2±0.4	1.9±0.1	1.6±0.4	2.2±0.5
caffeic acid euqivalent										
10	caffeic acid dimer	6.0±0.3	7.8±1.9	6.7±0.1	8.1±2.1	10.8±2.4	5.0±1.3	5.0±1.4	5.5±1.3	4.9±1.4
11	caffeic acid dimer	5.0±0.3	7.9±0.8	2.1±0.3	6.8±0.4	7.7±0.9	2.6±0.1	2.2±0.3	2.4±0.5	0.7±0.1
12	caffeic acid dimer	2.5±0.1	3.2±0.3	2.5±0.2	3.3±0.9	4.0±0.4	1.6±0.3	1.5±0.4	2.2±0.0	0.8±0.1
13	caffeic acid dimer	1.5±0.1	1.8±0.1	1.7±0.2	2.3±2.1	3.3±0.3	0.6±0.0	0.6±0.1	0.9±0.0	0.6±0.0
14	caffeic acid dimer	1.3±0.0	1.7±0.2	0.5±0.0	1.8±0.1	2.0±0.3	0.3±0.1	0.2±0.0	0.5±0.0	0.3±0.1
gallic acid euqivalent										
15	unknown	4.2±0.1	4.4±0.6	1.7±0.2	2.3±0.0	2.2±0.3	1.7±0.0	1.8±0.1	2.3±0.6	0.8±0.0
16	unknown	7.0±0.3	8.7±0.7	1.0±0.1	4.0±0.6	4.0±0.3	0.3±0.0	0.3±0.0	1.4±0.1	1.6±0.2
17	unknown	32.2±40.7	35.0±40.4	3.6±4.5	7.7±2.1	22.5±26.8	2.8±2.8	1.8±1.5	4.2±5.2	6.9±8.8
p-coumaric acid equivalent										
18	*p*-coumaric acid dimer	2.9±0.5	3.2±0.6	2.7±0.6	2.7±0.0	2.8±0.3	2.7±0.4	3.0±0.4	3.2±0.6	3.3±0.6
19	*p*-coumaric acid dimer	2.6±0.6	7.5±1.4	5.6±1.1	3.5±0.1	4.6±0.4	4.8±0.6	7.7±1.0	8.1±1.7	9.6±2.2
20	unknown	0.7±0.0	1.8±0.0	1.3±0.0	0.8±0.0	0.9±0.0	1.0±0.0	1.7±0.0	1.9±0.0	3.5±0.0
21	unknown	1.2±0.1	1.8±0.1	1.9±0.1	1.8±0.1	1.9±0.1	2.0±0.1	2.0±0.2	2.0±0.1	1.7±0.1
22	unknown	0.3±0.0	0.5±0.0	0.5±0.0	0.4±0.0	0.5±0.0	0.5±0.0	0.7±0.0	0.6±0.1	1.1±0.1
gallic acid euqivalent										
23	unknown	2.9±0.5	3.2±0.6	2.7±0.6	2.7±0.0	2.8±0.3	2.7±0.4	3.0±0.4	3.2±0.6	3.3±0.6
24	2,4,6-trihydroxy-benzaldehyde	48.9±0.4	19.4±1.9	41.5±1.6	65.1±4.0	29.4±6.1	36.1±0.9	58.7±6.3	45.0±4.3	42.0±15.4
25	anthocyanone A	17.1±1.8	14.6±0.1	20.5±1.5	23.8±2.6	28.7±4.7	21.5±0.1	18.5±2.7	15.5±1.9	33.4±0.4
26	syringic acid	25.9±4.2	34.5±3.5	35.5±0.6	21.8±3.8	51.0±9.7	41.5±3.7	18.8±0.9	28.5±5.5	36.4±7.1
27	vanillic acid	13.6±2.0	9.6±1.1	15.1±1.3	13.2±0.2	14.2±2.3	14.5±2.1	13.1±1.1	12.8±1.3	14.7±1.0

References

(1) Keller, M.; Viret, O.; Cole, F. M. *Botrytis cinerea* infection in grape flowers: defense reaction, latency, and disease expression. *Phytopathology* **2003**, 93, 316−322.

(2) Ky, I.; Lorrain, B.; Jourdes, M.; Pasquier, G.; Fermaud, M.; Gény, L.; Rey, P.; Doneche, B.; Teissedre, P. Assessment of grey mould (*Botrytis cinerea*) impact on phenolic and sensory quality of Bordeaux grapes, musts and wines for two consecutive vintages. *Aust. J. Grape Wine Res.* **2012**, 18, 215−226.

(3) Singleton, V. L.; Salgues, M.; Zaya, J.; Trousdale, E. Caftaric acid disappearance and conversion to products of enzymic oxidation in grape must and wine. *Am. J. Enol. Viticult.* **1985**, 36, 50−56.

(4) Cilliers, J. J. L.; Singleton, V. L. Caffeic acid autoxidation and the effects of thiols. *J. Agric. Food Chem.* **1990**, 38, 1789−1796.

(5) Cheynier, V.; Owe, C.; Rigaud, J. Oxidation of grape juice phenolic compounds in model solutions. *J. Food Sci.* **1988**, 53, 1729−1732.

(6) Cheynier, V. F.; Trousdale, E. K.; Singleton, V. L.; Salgues, M. J.; Wylde, R. Characterization of 2-*S*-glutathionyl caftaric acid and its hydrolysis in relation to grape wines. *J. Agric. Food Chem.* **1986**, 34, 217−221.

(7) Dubernet, M.; Ribéreau-Gayon, P. Les polyphénoloxydases du raisin sain et du raisin parasité par *Botrytis cinerea.* CR Acad. Sci. Paris 1973, 277, 975−978.

(8) Mayer, A. M.; Harel, E. Polyphenol oxidases in plants. *Phytochemistry* **1979**, 18, 193−215.

(9) Salgues, M.; Cheynier, V.; Gunata, Z.; Wylde, R. Oxidation of grape juice 2-*S*-glutathionyl caffeoyl tartaric acid by *Botrytis cinerea* laccase and characterization of a new substance: 2,5-di-*S*-glutathionyl caffeoyl tartaric acid. *J. Food Sci.* **1986**, 51, 1191−1194.

(10) Ribéreau-Gayon, P. New developments in wine microbiology. *Am. J. Enol. Viticult.* **1985**, 36, 1−10.

(11) Doneche, B. J. Botrytized wines. In *Wine microbiology and biotechnology*; Fleet, G. H., Ed.; Harwood Academic Publishers: Chur, Switzerland, 1993; Vol. 11, pp 327−353.

(12) Morales-Valle, H.; Silva, L. C.; Paterson, R. R.M.; Venâncio, A.; Lima, N. Effects of the origins of *Botrytis cinerea* on earthy aromas from grape broth media further inoculated with *Penicillium expansum*. *Food Microbiol.* **2011**, 28, 1048−1053.

(13) Grassin, C.; Dubourdieu, D. Quantitative determination of *Botrytis* laccase in musts and wines by the syringaldazine test. *J. Sci. Food Agric.* **1989**, 48, 369−376.

(14) Macheix, J.-J.; Sapis, J.-C.; Fleuriet, A.; Lee, C. Y. Phenolic compounds and polyphenoloxidase in relation to browning in grapes and wines. *Crit. Rev. Food Sci. Nutr.* **1991**, 30, 441−486.

(15) Perino, A.; Vercesi, A.; Fregoni, M. Confronto tra diverse metodiche utilizzate nella determinazione dell'infezione da *Botrytis cinerea* sui mosti. *Vignevini* **1994**, 7, 50−57.

(16) Kovac, V. Production de laccase par Botrytis cinerea sur milieu de culture artificiel. *Progr. Agric. Vitic.* **1982**, 99, 407−414.

(17) Zouari, N.; Romette, J.-L.; Thomas, D. Purification and properties of two laccase isoenzymes produced by *Botrytis cinerea. Appl. Biochem. Biotechnol.* **1987**, 15, 213−225.

(18) Slomczynski, D.; Nakas, J. P.; Tanenbaum, S. W. Production and characterization of laccase from *Botrytis cinerea* 61−34. *Appl. Environ. Microbiol.* **1995**, 61, 907−912.
(19) Vortkamp, A.; Rossetto, M.; Vanzani, P.; Di Paolo, M. L.; Klärner, S.; Muno-Bender, J.; Schneider, I.; Schnell, S.; Rigo, A.; Rauhut, D. Characterisation of laccase activity of two strains of *Botrytis cinerea*. *J. Plant Pathol.* **2013**, 95, 63−68.
(20) Dubernet, M.; Ribéreau-Gayon, P.; Lerner, H. R.; Harel, E.; Mayer, A. M. Purification and properties of laccase from *Botrytis cinerea*. *Phytochemistry* **1977**, 16, 191−193.
(21) Gigi, O.; Marbach, I.; Mayer, A. M. Induction of laccase formation in *Botrytis*. *Phytochemistry* **1980**, 19, 2273−2275.
(22) Gigi, O.; Marbach, I.; Mayer, A. M. Properties of gallic acid induced extracellular laccase of *Botrytis cinerea*. *Phytochemistry* **1981**, 20, 1211−1213.
(23) Marbach, I.; Harel, E.; Mayer, A. M. Inducer and culture medium dependent properties of extracellular laccase from *Botrytis cinerea*. *Phytochemistry* **1983**, 22, 1535−1538.
(24) Marbach, I.; Harel, E.; Mayer, A. M. Molecular properties of extracellular *Botrytis cinerea* laccase. *Phytochemistry* **1984**, 23, 2713−2717.
(25) Bradford, M. M. A rapid and sensitive method for the quantitation of microgram quantities of protein utilizing the principle of protein-dye binding. *Anal. Biochem.* **1976**, 72, 248−254.
(26) Harkin, J. M.; Obst, J. R. Syringaldazine, an effective reagent for detecting laccase and peroxidase in fungi. *Experientia* **1973**, 29, 381−387.
(27) Grindle, M. Phenotypic differences between natural and induced variants of *Botrytis cinerea*. *J. Gen. Microbiol.* **1979**, 111, 109−120.
(28) Lorenz, D. H. Analysis of morphological variation and pathogenesis in *Botrytis cinerea* Pers. and *Botrytinia fuckelinia* Whetz. *Z. Pflanzenkrankh. Pflanzenschutz* **1983**, 90, 622−633.
(29) Salinas, J.; Schot, C. P. Morphological and physiological aspects of *Botrytis cinerea*. *Mededelingen van de Rijksuniversiteit Faculteit Landbouwwetenschappen, Gent* **1987**, 52, 771−776.
(30) Leone, G. In vivo and in vitro phosphate-dependent polygalacturonase production by different isolates of *Botrytis cinerea*. *Mycol. Res.* **1990**, 94, 1039−1045.
(31) Movahedi, S.; Heale, J. B. Purification and characterization of an aspartic proteinase secreted by *Botrytis cinerea* Pers ex. Pers in culture and in infected carrots. *Physiol. Mol. Plant Pathol.* **1990**, 36, 289−302.
(32) Xu, F.; Berka, R. M.; Wahleithner, J. A.; Nelson, B. A.; Shuster, J. R.; Brown, S. H.; Palmer, A. E.; Solomon, E. I. Site-directed mutations in fungal laccase. Effect on redox potential, activity and pH profile. *Biochem. J.* **1998**, 334, 63−70.
(33) Madzak, C.; Mimmi, M. C.; Caminade, E.; Brault, A.; Baumberger, S.; Briozzo, P.; Mougin, C.; Jolivalt, C. Shifting the optimal pH of activity for a laccase from the fungus *Trametes versicolor* by structure-based mutagenesis. *Protein Eng., Des. Sel.* **2006**, 19, 77−84.
(34) Zumárraga, M.; Camarero, S.; Shleev, S.; Martínez-Arias, A.; Ballesteros, A.; Plou, F. J.; Alcalde, M. Altering the laccase functionality by *in vivo* assembly of mutant libraries with different mutational spectra. *Proteins: Struct., Funct., Genet.* **2008**, 71, 250−260.

(35) Viterbo, A.; Yagen, B.; Mayer, A. M. Induction of laccase formation in *Botrytis* and its inhibition by cucurbitacin: is gallic acid the true inducer? *Phytochemistry* **1993**, 34, 47−49.
(36) Leonowicz, A.; Trojanowski, J.; Orlicz, B. Induction of laccase in Basidiomycetes: apparent activity of the inducible and constitutive forms of the enzyme with phenolic substrates. *Acta Biochim. Polym.* **1978**, 25, 369−378.
(37) Bollag, J.-M.; Leonowicz, A. Comparative studies of extracellular fungal laccases. *Appl. Environ. Microbiol.* **1984**, 48, 849−854.
(38) Van Rensburg, P.; Pretorius, I. S. Enzymes in winemaking: harnessing natural catalysts for efficient biotransformations-A review. *S. Afr. J. Enol. Vitic.* **2000**, 21, 52−73.
(39) Vivas, N.; Vivas de Gaulejac, N. V.; Vitry, C.; Bourden-Nonier, M. F.; Chauvet, S.; Doneche, B.; Absalon, C.; Mouche, C. Occurrence and specificity of glucose oxidase (EC. 1.1. 3.4) in botrytized sweet white wine. Comparison with laccase (EC: 1.10. 3.2), considered as the main responsible factor for oxidation in this type of wine. *Vitis* **2010**, 49, 113−120.
(40) Guyot, S.; Cheynier, V.; Souquet, J.-M.; Moutounet, M. Influence of pH on the enzymatic oxidation of (+)-catechin in model systems. *J. Agric. Food Chem.* **1995**, 43, 2458−2462.
(41) Guyot, S.; Vercauteren, J.; Cheynier, V. Structural determination of colourless and yellow dimers resulting from (+)-catechin coupling catalysed by grape polyphenoloxidase. *Phytochemistry* **1996**, 42, 1279−1288.
(42) Weinges, K.; Huthwelker, D. Oxydative Kupplung von Phenolen, III1) Isolierung und Konstitutionsbeweis eines 8.6'- verknüpften Dehydro-dicatechins (B4). *Justus Liebigs Ann. Chem.* **1970**, 731, 161−170.
(43) Weinges, K.; Ebert, W.; Huthwelker, D.; Mattauch, H.; Perner, J. Oxydative Kupplung von Phenolen, II1) Konstitution und Bildungsmechanismus des Dehydro-dicatechins A. *Justus Liebigs Ann. Chem.* **1969**, 726, 114−124.
(44) Weinges, K.; Mattauch, H. Der chemische konstitutionsbeweis des dehydro-dicatechins A. *Chemik. Zeit.* **1971**, 95, 155.
(45) Carunchio, F.; Crescenzi, C.; Girelli, A. M.; Messina, A.; Tarola, A. M. Oxidation of ferulic acid by laccase. Identification of the products and inhibitory effects of some dipeptides. *Talanta* **2001**, 55, 189−200.
(46) Smirnov, S. A.; Koroleva, O. V.; Gavrilova, V. P.; Belova, A. B.; Klyachko, N. L. Laccases from basidiomycetes. Physicochemical characteristics and substrate specificity towards methoxyphenolic compounds. *Biochemistry (Moscow)* **2001**, 66, 774−779.
(47) D'annibale, A.; Celletti, D.; Felici, M.; Di Mattia, E.; Giovannozzi-Sermanni, G. Substrate specificity of laccase from *Lentinus edodes. Acta Biotechnol.* **1996**, 16, 257−270.
(48) Feng, X. Oxidation of phenols, anilines, and benzenethiols by fungal laccases. Correlation between activity and redox potentials as well as halide inhibition. *Biochemistry* **1996**, 35, 7608−7614.
(49) Janovitz-Klapp, A. H.; Richard, F. C.; Goupy, P. M.; Nicolas, J. J. Inhibition studies on apple polyphenol oxidase. *J. Agric. Food Chem.* **1990**, 38, 926−931.
(50) Son, S. M.; Moon, K. D.; Lee, C. Y. Inhibitory effects of various antibrowning agents on apple slices. *Food Chem.* **2001**, 73, 23−30.

(51) Shengzhao, G.; Jiang, C.; Zhuoru, Y. Inhibitory effect of ferulic acid on oxidation of L-DOPA catalyzed by mushroom tyrosinase. *Chin. J. Chem. Eng.* **2005**, 13, 771−775.
(52) Božič, M.; Gorgieva, S.; Kokol, V. Laccase-mediated functionalization of chitosan by caffeic and gallic acids for modulating antioxidant and antimicrobial properties. *Carbohydr. Polym.* **2012**, 87, 2388−2398.
(53) Tulyathan, V.; Boulton, R. B.; Singleton, V. L. Oxygen uptake by gallic acid as a model for similar reactions in wines. *J. Agric. Food Chem.* **1989**, 37, 844−849.
(54) Mathew, A. G.; Parpia, H. A. Food browning as a polyphenol reaction. *Adv. Food Res.* **1971**, 19, 75−145.
(55) Lopes, P.; Richard, T.; Saucier, C.; Teissedre, P.-L.; Monti, J.-P.; Glories, Y. Anthocyanone A. A quinone methide derivative resulting from malvidin 3-*O*-glucoside degradation. *J. Agric. Food Chem.* **2007**, 55, 2698−2704.
(56) Leonowicz, A.; Edgehill, R. U.; Bollag, J.-M. The effect of pH on the transformation of syringic and vanillic acids by the laccases of *Rhizoctonia praticola* and *Trametes versicolor*. *Arch. Microbiol.* **1984**, 137, 89−96.

Chapter 4

Concluding remarks

1 Interactions of copper and glutathione during fermentation

The threat of fungal infections in the vineyard represents a worldwide problem in viticulture and enology. The susceptibility of *Vitis vinifera* L. to fungal infections, the ubiquitousness of fungal pathogens, and their strain-specific variability make the disease management a great challenge. A vast spectrum of synthetic fungicides exists, however, because of increasing concerns on their effect on environmental pollution and human health, organic wine production has increased over the last 10 years (Pedneault & Provost, 2016). Copper-based fungicides, such as the Bordeaux mixture and copper sulfate, are the most commonly used fungicides in organic viticulture (Cavazza et al., 2013). Must and wine from regions which highly suffer from fungal infections due to climatic conditions have a higher risk to contain elevated copper levels. It is well established that a high reactivity of copper with wine components deteriorates organoleptic wine quality (Li et al., 2008; Ugliano et al., 2011). This is manifested in a decrease in important aroma compounds and discoloration of the wine (see **Chapter 1**, Section 2.3). Further problems occur due to the toxicity of high copper concentrations toward the wine yeast *Saccharomyces cerevisiae*. Above the concentrations that are essential for yeast metabolism, this leads to a loss of membrane integrity and an inhibition of important enzyme activities (Ohsumi et al., 1988; Shanmuganathan et al., 2004). The knowledge on glutathione (GSH) to be an important redox-active constituent of the yeast and its participation in the detoxification of heavy metals (Penninckx, 2000) led to the assumption that GSH serves as an optimal additive to the must to reduce copper stress.

Results of the present work showed that the vitality of a commercially available yeast strain was considerably reduced at copper concentrations ranging from 10 to 25 mg·L^{-1} (see **Chapter 2**). The same concentrations resulted in a compromised fermentation efficiency of the yeast with an initially decelerated sugar degradation and lower rates of maximum sugar degradation. This increases the risk of a stuck and sluggish fermentation. Such uncontrolled fermentation hinders the principal objective of winemakers to achieve the desired percentage of sugar transformation to CO_2 and alcohol. Resulting wines may also have an increased risk of bacterial spoilage due to high sugar levels (Ribéreau-Gayon et al., 2006). However, the observed effects of different copper concentrations on yeasts were strain-dependent. Even the highest copper concentration (25 mg·L^{-1}) did not influence yeast vitality and fermentation efficiency of the yeast Anaferm Riesling.

Analysis of copper concentrations in the must (Cu_{ex}) and intracellular copper concentrations in the enzyme extracts (Cu_{in}) demonstrated an important difference between the copper-sensitive yeast strain Oenoferm X-treme and the resistant strain Anaferm Riesling. Copper resistance of the yeast Anaferm Riesling might be a reason for the good copper adsorption to the cell surface, combined with a reduced copper influx. This is implicated by a faster decrease in Cu_{ex} concentrations and lower Cu_{in} concentrations compared to the copper-sensitive yeast strain. According to this, the present work proved that high removal efficiency of copper from extracellular media is not necessarily associated with high copper transport into the cells. The general decrease in Cu_{ex} values for both yeasts is likely due to the extracellular adsorption and import of copper by the yeast to maintain copper homeostasis (Vest et al., 2019). This is of great importance regarding copper as an essential micronutrient for metabolic processes. Transport systems at the plasma membrane have been described for the copper import (Culotta et al., 1999). If copper reaches toxic levels, the yeast is able to limit the toxic influx and react with different copper detoxification mechanisms such as metallothioneins and GSH (see **Chapter 1**, Section 2.2 and 4.4). This ability might be more pronounced by the yeast Anaferm Riesling than by the yeast Oenoferm X-treme, resulting in a higher copper resistance.

The addition of 20 mg·L^{-1} GSH to the must was suitable to reduce the effects of copper on the copper-sensitive yeast strain at concentrations ≥10 mg·L^{-1}. The yeast vitality was considerably increased and a positive influence on fermentation efficiency was observed by a distinct reduction of the lag phase of sugar degradation and an increase in the maximum rate of sugar degradation. Besides the redox activity and the protection against reactive oxygen species, GSH has the ability to complex heavy metals (Połeć-Pawlak et al., 2007). This led to the assumption that GSH might complex copper in the must, which consequently reduces the toxic effects of free copper ions on the yeast cells. Complexation might occur via the thiol group to form a mercaptide bond resulting in a Cu-GSH complex (Corazza et al., 1996). Another possibility is that copper oxidizes GSH to glutathione disulfide (GSSG) and further forms a Cu-GSSG complex due to the high affinity of copper to the amine and carboxylic group (Połeć-Pawlak et al., 2007). It is also likely that GSH exhibits an intracellular protective effect against copper. The content of the intracellularly synthesized GSH by the yeast might be increased by assimilation of the GSH added to the must via specific membrane transport proteins (see **Chapter 1**, Section 4.4.1).

Analysis of the GSH content in the must showed a low recovery of the GSH added, even in samples immediately analyzed after the addition of GSH. This confirms that it directly participated in reactions in the must or was assimilated by the yeast to serve as a stress response factor. GSH concentrations decreased inversely proportional to the copper concentrations, giving additional evidence for the complexation of copper and GSH, or increased oxidation of GSH to its disulfidic form. The significantly higher availability of GSH in microvinifications (MVs) with GSH addition might be the reason for an enhanced protection of the yeast against excess copper. In MV trials with the copper-resistant yeast strain, GSH was not detectable in the must until day 23 and GSH addition resulted in generally lower GSH concentrations in the must, compared to the copper-sensitive yeast. GSH was presumably more efficiently utilized by the yeast Anaferm Riesling for copper detoxification, which consequently led to copper resistance. During fermentation, an increase in extracellular GSH was observed for both yeasts and might be associated with yeast autolysis as it has been reported in a previous study (Dubourdieu & Lavigne-Cruege, 2004). Further analysis of extracellular Cu-GSH or Cu-GSSG complexes and intracellular GSH contents might give additional information where the main place of interaction occurs. The comparison of intra- and extracellular GSH would further help to elucidate the process of GSH secretion and assimilation by the yeast. However, it would remain unknown, whether the intracellular GSH emerge from the import or synthesis by the yeast. The use of labeled GSH (^{35}S) could reveal suitable information on the pathways of GSH (Kritzinger et al., 2012).

The interaction of copper and GSH during fermentation was further proven by the comparison of intra- and extracellular copper concentrations of MV trials with copper as the sole additive and in combination with GSH. The improved yeast vitality and shorter adaption phase of the copper-sensitive yeast strain by GSH addition presumably resulted in an enhanced proliferation. This was reflected in a faster initial decrease of Cu_{ex} concentrations, compared to MVs without GSH addition. The comparison of Cu_{ex} concentrations in musts with and without GSH addition showed considerably higher Cu_{ex} concentrations without GSH addition for both yeasts, but the difference of both concentrations was smaller for the copper-resistant strain. The reduction of extracellular copper due to extracellular complexation with GSH can be excluded because AAS analysis determined the total Cu_{ex} content in the must, including complexes with GSH. GSH consequently enhances copper biosorption and the effect was higher for the copper-sensitive strain, indicating that the effect is also dependent on the specific yeast strain properties. In contrast, Cu_{in} values were generally lower in musts with GSH addition at high copper concentrations, compared to those without GSH addition. This indicates a lower copper influx due to GSH addition. Increased Cu_{in} concentrations, however, do not necessarily implicate a higher toxicity on cells, since intracellular copper is not present in its free form. It is bound to chaperones, metallothioneins, or GSH and can be efficiently sequestered by its transport into organelles such as vacuoles to prevent oxidative stress or inappropriate binding interactions (Vest et al., 2019). The important role of GSH in the copper transfer to protective enzymes such as metallothioneins after its complexation has been reported (Ferreira et al., 1993). The detoxification mechanism of copper by yeasts is very complex and is not only driven by GSH. Interactions of several proteins are of great importance if copper reaches toxic levels. It would be interesting to analyze the contribution of this inherent protection mechanism of the yeast in different MV trials. The sole determination of copper and GSH concentrations, thus, is only one step to elucidate the complex interaction of GSH and copper during fermentation and to evaluate the impact of GSH on copper stress. The occurrence of such an impact was clearly demonstrated in the present work by significant differences of several parameters in MV trials with and without GSH addition.

Analysis of acetaldehyde concentrations revealed a further parameter that is distinctively influenced by high copper concentrations in the must and GSH addition. The temporary acetaldehyde accumulation increased proportionally to the copper concentrations and was more pronounced for the copper-sensitive yeast strain, compared to the copper-resistant strain. GSH addition distinctively mitigated acetaldehyde accumulation at copper concentrations $\geq$10 mg·L^{-1}. Consequently, acetaldehyde accumulation might be used as an indicator of copper stress in the yeast, which can be reduced by GSH addition. High acetaldehyde concentrations are undesired in winemaking since its high reactivity may deteriorate wine aroma and color and reduces the functionality of SO_2 (see **Chapter 1**, Section 2.3). The possible toxicity of acetaldehyde against the yeast, which might cause the risk of a reduced cell population, glucose metabolism, and ethanol production can be neglected. Maximum acetaldehyde concentrations detected in the present work are still below the reported level that influenced yeast growth and fermentation efficiency (Maiorella et al., 1983). The detection of high acetaldehyde concentrations led to another aim of this work: the analysis of specific alcohol dehydrogenase (ADH) activity. Results implied that this enzyme is affected by copper since it catalyzes the reaction from acetaldehyde to ethanol and vice versa. It should further serve as an example for the determination of the influence of copper alone and in combination with GSH on enzyme activities. It is important to consider enzyme activities in this experiment because proteins represent the major target of oxidation by copper (Shanmuganathan et al., 2004). The results of previous studies confirmed that the ADH is a representative enzyme to reflect the influence of copper on enzymes. The isoenzyme ADHI has strongly been affected by both activities, reducing and inhibiting mechanisms, which are described for copper (see **Chapter 1**, Section 2.1). On the one hand, it has been shown to represent the most severely affected enzyme by copper oxidation, compared to several catabolic enzymes (Shanmuganathan et al., 2004). On the other hand, the substitution of the catalytic zinc at the active site by copper has been reported to reduce specific ADH activity (Cavaletto et al., 2000; Vanni et al., 2002). In the present work, the specific ADH activity was observed after 1 week of fermentation. Results confirmed a copper-mediated inhibition or reduction of enzyme activities. Reductions increased proportionally to copper concentrations, reaching no detectable activity for the copper-sensitive strain at a copper exposition of ≥10 mg·L^{-1} and for the copper-resistant strain at 25 mg·L^{-1} copper. Total inhibition can be excluded at these concentrations, since an *in vivo* ADH activity can be deduced from a decrease in acetaldehyde in all MVs. The work further revealed an ambivalent influence of GSH addition on the ADH activity of the copper-sensitive strain. This was again implicated in the interaction of copper and GSH. On the one hand, GSH addition restored ADH activity at a high copper concentration of 10 mg·L^{-1}, presumably due to the reduced copper exposition after the complexation of copper and GSH. On the other hand, GSH addition lowered ADH activity in the low copper trials (control and 2.5 mg·L^{-1}), which were below the stress-inducing level of the copper-sensitive yeast. It is assumed that the complexation of extracellular copper by GSH resulted in a copper deficiency, negatively influencing enzymatic processes in the yeast. Yeast cells require low amounts of copper as an essential micronutrient for metabolic processes (see **Chapter 1**, Section 2.1).

The present work revealed new insights in the influence of pure GSH addition on fermentations with copper-rich musts and the interactions of both compounds. It is an important step to determine all factors which might influence GSH addition to must and wine to obtain more reliability in its application in winemaking. One of these important factors is copper, due to its high reactivity with wine components and its ambivalence of essentiality and toxicity for the yeast. The addition of 20 mg·L^{-1} GSH was suitable to reduce the negative effects of copper on a copper-sensitive yeast strain. However, results also showed that the knowledge on the initial copper concentration in the must and the consecutive choice of a suitable yeast strain is of great importance in winemaking. This can help to avoid stressful conditions for the yeast during fermentation and, consequently, wine quality deterioration. The presented results further demonstrate the difficulty in the search for an adequate limitation of copper concentrations in the must. The permitted limit of ≤20 mg·L^{-1} in the European Union (Ferreira et al., 2006) might be suitable for the yeast Anaferm Riesling, however, the fermentation efficiency of the yeast Oenoferm X-treme is already influenced at half of the concentration.

2 Impact of laccases from different *Botrytis cinerea* isolates on wine phenolic compounds

Botrytis cinerea not only indirectly accounts for wine quality deterioration due to its control with copper-based fungicides. As a consequence of inappropriate or limited disease management, *B. cinerea* may spread in the vineyard. Once the fungus infects grape berries, the secretion of several enzymes starts to counteract the host-defense mechanisms. Among these enzymes, laccase is responsible for the oxidation of a wide range of substrates, mainly phenolic compounds, produced as antifungal agents in the grape (Ky et al., 2012). This natural defense mechanism creates a great threat of quality deterioration during the process of winemaking. If laccase is present in the must after processing of infected grapes, the oxidation of phenols results in the formation of brown products that negatively influence wine color (Salgues et al., 1986). This reaction also occurs in musts prepared from non-infected grapes due to the presence of tyrosinase. However, this enzyme can easily be controlled by SO_2 addition or cooling, whereas laccase is very resistant to the common inactivation procedures and is more stable at wine conditions (Dewey et al., 2008; Ugliano, 2009). Additional problems occur due to the wider substrate spectrum of laccase, which is even able to further oxidize the grape reaction product (GRP) (see **Chapter 1**, Section 4.3.1) (Salgues et al., 1986). *Botrytis* infestation of the grape cannot completely be avoided and, consequently, most musts and wines contain certain amounts of laccase. In order to face this problem, it is important to evaluate the catalytic activity of the enzyme toward typical wine phenols. This might help to further assess the deterioration capacity of laccases on wine organoleptic properties. Previous studies have demonstrated that methods like the sole determination of the degree of infestation on the grapes are insufficient to estimate laccase activity. A high degree of infection does not necessarily result in a high laccase activity in the must (Macheix et al., 1991; Perino et al., 1994). Thus, the aim of the present work was to analyze catalytic activities of laccases of various *B. cinerea* isolates from different grapes varieties (see **Chapter 3**).

Laccase production was induced for 3 days in a buffered malt extract medium in the presence of gallic acid, which has been described to be an efficient inductor for laccase production (Gigi et al., 1980). This method exploits the natural defense mechanism of the fungus, to produce and secrete laccase for the degradation and inactivation of antifungal phenolic compounds. Laccase and other pathogen-related extracellular enzymes are released in a glucan-rich secretome by the fungus, which has been suggested to further enhance its resistance to host responses (Gil-ad et al., 2001). Purification of laccase from the secretomes was not conducted since the assays were carried out in a buffer solution which did not contain substrates for other enzymes except those for laccase. Commercially available *Trametes versicolor* laccase was used for comparison to the isolates in all tests. Syringaldazine served as a laccase-specific substrate to determine and compare specific laccase activities in the secretomes of different isolates. This assay proved the presence of laccase in the secretome with a large variability of specific activities ranging from 0.03 to 2.33 $U_{Lacc} \cdot mg^{-1}$. The significantly highest laccase activity was ascribed to an isolate from Pinot blanc grapes, whereas the lowest activities were determined for two isolates from Pinot noir grapes. The two isolates were not applied in further assays due to the limited amount of protein in the extract. No similarities in the specific activities of isolates from the same grape variety were observed. The specific activity of *T. versicolor* laccase was in a comparatively low range with 0.13 $U_{Lacc} \cdot mg^{-1}$ despite its high purity.

The next step was to evaluate the catalytic activity of different laccase-containing secretomes toward six wine phenols by the determination of substrate half-life values. According to the syringaldazine assay, all laccases were applied with 0.001 U_{Lacc}. As expected, the substrates caffeic and ferulic acids showed the shortest substrate half-life values for all isolates. High specificities for both substrates have already been observed for *B. cinerea* laccases (Dubernet et al., 1977; Marbach et al., 1984), and hydroxycinnamic acids have been described to be a major class of phenolic compounds in white must and wine participating in enzymatic browning (Li et al., 2008). The high substrate affinity can be ascribed to specific substrate structures, e.g., the methoxy group of ferulic acid, which stabilizes the radical formed after oxidation and increases the catalytic effectiveness (Smirnov et al., 2001; D'annibale et al., 1996; Feng, 1996). The most prevalent anthocyanin in red wines, malvidin 3-*O*-glucoside, also showed short substrate half-life values. This high substrate affinity might cause a risk for red wine quality since oxidation resulted in a loss of the red color. The lowest half-life values determined for (+)-catechin, gallic acid, and *p*-coumaric acid were even above the highest values observed for the other phenols. This might be deduced from structural differences and indicates a lower substrate affinity. This lowers the risk of color deterioration compared to the other substrates. In case of gallic acid, the third hydroxyl group might decelerate oxidation by laccases. By averaging the half-life values of all isolates for one substrate, the following order can be determined with an increasing duration of substrate oxidation: caffeic acid < ferulic acid < malvidin 3-*O*-glucoside < (+)-catechin < gallic acid < *p*-coumaric acid. No correlations between specific activities determined with the syringaldazine assay and substrate half-life values were observed. Isolates with a high activity toward syringaldazine do not necessarily show faster substrate oxidation compared to the other isolates and isolates from the same grape variety did not lead to similar substrate half-life values for particular substrates.

Detailed information on the formation of oxidation products is of great concern, since products are responsible for the undesired discoloration of must and wine, mainly as a result of condensation and polymerization reactions. Reaction mixtures turned to a yellow, greenish-yellow, or brown color during oxidation of (+)-catechin, gallic acid, and caffeic acid, respectively. Product profiles were revealed for different substrates. It has to be considered that the formation of products depends on the catalytic activity of laccases, whereas the decrease in relative product concentrations is enzyme-independent due to the high reactivity of the quinones formed. For (+)-catechin, ferulic, caffeic, and *p*-coumaric acids, oxidation products were mainly identified as substrate dimers. For ferulic acid, a trimer was additionally identified. Product identification for gallic acid was not possible and for two strains isolated from Pinot noir and Portugieser grapes one of three products was not detected. The greenish-brown color of the reaction mixture was indicative of the presence of products with a high degree of polymerization, which limited the detection of distinct product peaks. The color change in the reaction mixtures with (+)-catechin can presumably be ascribed to the formation of yellow-colored dehydrocatechins A. Absorption maxima of three products correspond to yellow derivatives, which has been previously described after (+)-catechin oxidation by a crude grape polyphenoloxidase (Guyot et al., 1995; Guyot et al., 1996). One of these products was tentatively identified as dehydrocatechin A. Reaction mixtures with oxidation products of caffeic acid appeared in a brownish color, which has been described to be characteristic for caffeic acid polymers (Cilliers & Singleton, 1990). Reaction mixtures of ferulic and *p*-coumaric acids did not show a color change after oxidation. These reactions, however, are as well important from a quality point of view when they take place in must and wine. The reactive quinones formed may further react with other phenols or other wine components leading to color deterioration or oxidation of important aroma compounds (Li et al., 2008; Ugliano et al., 2011). Oxidation of malvidin 3-*O*-glucoside differed from other substrates since it led to a loss of the red color in the reaction mixture. It was shown that oxidation products emerge from oxidative degradation of the substrate, while those of other substrates occurred after polymerization. Cleavage of the B-ring of malvidin 3-*O*-glucoside resulted in the identified product syringic acid, while the product vanillic acid corresponds to the B-ring moiety of peonidin 3-*O*-glucoside, which was present in small amounts in the standard used. Further products were identified as 2,4,6-trihydroxybenzaldehyde and anthocyanone A (Lopes et al., 2007), a nonspecific degradation product, which might be formed by oxidation of both anthocyanins. Laccase activities of different strains and the subsequent non-enzymatic oxidation led to mainly the same product profiles for the six substrates, indicating that the negative discoloration might be the same if different laccases are present in must or wine. It depends much more on the chemical composition of the wine, including phenolic and non-phenolic compounds, and the following chemical oxidation, which is not influenced by laccases after the first enzyme-catalyzed step.

Concerning different half-life values for specific substrates, it was assumed that the extent of discoloration might differ between laccases. The rate of product formation and maximum relative product concentrations were considered as additional parameters. This might help to estimate how fast discoloration of musts may occur after oxidation of phenols by different secretomes. Regarding the progress of product formation, results could demonstrate that for some substrates larger variations between laccases occur, while for other substrates the progress was nearly similar. It was further shown that the chemical oxidation after the laccase-catalyzed step differs between substrates, depending on the stability of products. But the reaction, even if it is enzyme-independent, is connected to the first enzyme-catalyzed step. This was shown by differences in the decrease in the same products for different isolates, and thus, an indirect dependence is apparent. In the following, considerable distinctions and similarities between different isolates for different phenols are described. The short substrate half-life values for ferulic and caffeic acids correspond to a fast increase in relative product concentrations, reaching maximum product concentrations after 1–4 h, depending on the different secretomes. The subsequent decrease in products occurred quickly, with a total depletion of products within 24–48 h. The sole exception was observed for an isolate from Riesling grapes. Compared to the other isolates, it showed maximum relative product concentrations for ferulic acid after 24 h and a higher product stability because products were still present in higher concentrations after 48 h. One predominant dimer was observed for all secretomes after oxidation of caffeic and ferulic acids, but the maximum relative concentration of this ferulic and caffeic acid dimer differed between isolates. In contrast, for (+)-catechin, differences were more pronounced between isolates. Oxidation of this usually most abundant flavanol in wine is of great importance for wine color, since the predominant products identified in this work correspond to yellow derivatives, mainly the yellow dehydrocatechin A. Maximum product concentrations of different secretomes were reached within a greater range of time (4–24 h), compared to ferulic and caffeic acids. The predominant product differed for two isolates from Pinot blanc grapes. It seemed to be less stable compared to the product of other isolates since the decrease led to a complete diminution after 48 h. Relative product concentrations after gallic acid oxidation were mainly low, which was attributed to low stability of the products. Only one compound was observed in high concentrations for two Riesling and one Pinot blanc isolate. Since product identification was not possible, an impact on color change cannot be deduced. For *p*-coumaric acid, a slow increase in product concentrations corresponds to the long half-life for this substrate. Maximum product concentrations were observed after 24–48 h. Only for an isolate from Riesling grapes and two isolates from Pinot noir grapes one product reached its maximum concentration after 4 h and the predominant product was determined in higher concentrations, compared to the other isolates. Oxidation of malvidin 3-*O*-glucoside by laccases relies on oxidative cleavage. Consequently, the products differ from those of other substrates, which mainly formed dimers. The change of relative product concentration over time indicated high product stability in contrast to products from ferulic and caffeic acids, which showed similar substrate half-life values. The product increase for malvidin 3-*O*-glucoside occurred slowly until 24–48 h of oxidation and all products were detected in relatively high concentrations. The predominant products were syringic acid or trihydroxybenzaldehyde depending on the different secretomes. One isolate from Pinot noir grapes, which had the shortest substrate half-life, differed from others, reaching maximum product concentrations of syringic acid and an unknown product already after 4 h. The results of this work clearly demonstrate a varying catalytic activity of laccases from different isolates toward typical wine phenols. It was not possible to deduce whether one particular *B. cinerea* isolate or all isolates from a particular grape variety are responsible for faster laccase-mediated substrate oxidation or faster product formation rates. For each substrate, laccases of different isolates led to distinctively shorter substrate half-life values, higher maximum product concentrations, faster rates of product formation and decrease, and differing predominant products. These properties account for laccases with a high risk of color deterioration. In contrast, the presence of laccases in musts with long substrate half-life values and a delayed product formation give winemakers a better chance to reduce oxidative color deterioration, if procedures described to reduce laccase activity are applied quickly (see **Chapter 1**, Section 3.2). The results of the present work may help to estimate the time within the procedures are still effective. For example, the risk of yellow discoloration after (+)-catechin oxidation seems to be comparatively low for musts containing laccase of the Pinot blanc isolate showing a half-life of 16.1 h, while a higher risk occurs with the same amount of a laccase from a Pinot noir isolate having a half-life of 6.6 h. This estimation of a higher risk is reinforced with a much faster production of maximum product concentrations within 4 h, compared to 24 h for the Pinot blanc isolate. In contrast, brown discoloration due to caffeic acid oxidation with laccases from the same Pinot noir and Pinot blanc isolates is more likely in musts with both laccases, showing substrate half-life values of 0.4 and 1.8 h, respectively. The shorter half-life for the Pinot noir isolate is accompanied by a faster maximum product formation within 1 h instead of 4 h for the Pinot blanc isolate.

It needs to be considered that the present work only revealed reactions in buffer solutions. As described above, must and wine represent complex matrices, where several components can participate in and influence oxidation. In this context, GSH and copper need to be mentioned again. GSH is able to reduce the browning capacity of wines, which has been evaluated by a low hydroxycinnamic acid-to-GSH ratio (Du Toit et al., 2006), whereas copper enhances oxidative browning of wine (Li et al., 2008). On the other hand, copper has been described as an efficient laccase inhibitor (Lorenzo et al., 2005). It would be interesting to further apply GSH and copper in this experiment to analyze its influence and interactions on different laccases and the products of specific substrate oxidations. Laccases might not only affect wine color by the oxidation of phenolic compounds, but also the mouthfeel of the wine and its beneficial antioxidant potential might be influenced. In terms of taste and mouthfeel, for example, catechin is relevant to the quality of red wines as a key compound for astringency, and gallic acid, a precursor of hydrolyzable tannins, is described to be responsible for some bitterness and astringency (Claus et al., 2014). The deterioration of the antioxidant capacity of wines was shown to be a result of oxidative browning since oxidized phenols might lose their ability to serve as hydrogen donors and reducing agents (Sioumis et al., 2005). Evaluation of the influence of different laccases on these parameters would further be an important research topic. Once again, GSH needs to be considered as a potent natural additive in winemaking. It can limit the process of polymerization by trapping reactive products emerging from laccase-catalyzed oxidation of wine phenols. This might lead to a preservation of wine color, aroma, mouthfeel, and nutritional value.

References

Cavaletto, M.; Pessione, E.; Vanni, A.; Giunta, C. Improved resistance to transition metals of a cobalt-substituted alcohol dehydrogenase 1 from *Saccharomyces cerevisiae*. *Journal of Biotechnology*, **2000**, 84, 87–91.

Cavazza, A.; Guzzon, R.; Malacarne, M.; Larcher, R. The influence of the copper content in grape must on alcoholic fermentation kinetics and wine quality: a survey on the performance of 50 commercial active dry yeasts. *Vitis*, **2013**, 52, 149–155.

Cilliers, J. J. L.; Singleton, V. L. Caffeic acid autoxidation and the effects of thiols. *Journal of Agricultural and Food Chemistry*, **1990**, 38, 1789–1796.

Claus, H.; Sabel, A.; König, H. Wine phenols and laccase: An ambivalent relationship. In *Wine phenolic composition, classification and health benefits*; El Rayess, Y., Ed.; Nova Publishers: New York, USA, 2014, 155–185.

Corazza, A.; Harvey, I.; Sadler, P. J. ^{1}H, ^{13}C-NMR and X-ray absorption studies of copper (I) glutathione complexes. *European Journal of Biochemistry*, **1996**, 236, 697–705.

Culotta, V. C.; Lin, S.-J.; Schmidt, P.; Klomp, L. W. J.; Casareno, R. L. B.; Gitlin, J. Intracellular pathways of copper trafficking in yeast and humans. In *Copper transport and its disorders, Molecular and cellular aspects*; Leone, A.; Mercer, J. F. B., Eds.; Kluwer Academic / Plenum Publisher: New York, USA, 1999, 448, 247–254.

D'annibale, A.; Celletti, D.; Felici, M.; Di Mattia, E.; Giovannozzi-Sermanni, G. Substrate specificity of laccase from *Lentinus edodes*. *Acta Biotechnologica*, **1996**, 16, 257–270.

Dewey, F. M.; Hill, M.; DeScenzo, R. Quantification of *Botrytis* and laccase in winegrapes. *American Journal of Enology and Viticulture*, **2008**, 59, 47–54.

Du Toit, W. J.; Marais, J.; Pretorius, I. S.; Du Toit, M. Oxygen in must and wine: A review. *South African Journal of Enology and Viticulture*, **2006**, 27, 76–94.

Dubernet, M.; Ribereau-Gayon, P.; Lerner, H. R.; Harel, E.; Mayer, A. M. Purification and properties of laccase from *Botrytis cinerea*. *Phytochemistry*, **1977**, 16, 191–193.

Dubourdieu, D.; Lavigne-Cruege, V. The role of glutathione on the aromatic evolution of dry white wine. *Vinidea.net Wine Internet Technical Journal*, **2004**, 2, 1–9.

Feng, X. Oxidation of phenols, anilines, and benzenethiols by fungal laccases. Correlation between activity and redox potentials as well as halide inhibition. *Biochemistry*, **1996**, 35, 7608–7614.

Ferreira, D. C.; Ciriolo, M. R.; Marcocci, L.; Rotilio, G. Copper (I) transfer into metallothionein mediated by glutathione. *Biochemical Journal*, **1993**, 292, 673–676.

Ferreira, J.; Du Toit, M.; Du Toit, W. J. The effects of copper and high sugar concentrations on growth, fermentation efficiency and volatile acidity production of different commercial wine yeast strains. *Australian Journal of Grape and Wine Research*, **2006**, 12, 50–56.

Gigi, O.; Marbach, I.; Mayer, A. M. Induction of laccase formation in *Botrytis*. *Phytochemistry*, **1980**, 19, 2273–2275.

Gil-ad, N. L.; Bar-Nun, N.; Mayer, A. M. The possible function of the glucan sheath of *Botrytis cinerea*: effects on the distribution of enzyme activities. *FEMS Microbiology Letters*, **2001**, 199, 109–113.

Guyot, S.; Cheynier, V.; Souquet, J.-M.; Moutounet, M. Influence of pH on the enzymatic oxidation of (+)-catechin in model systems. *Journal of Agricultural and Food Chemistry*, **1995**, 43, 2458–2462.

Guyot, S.; Vercauteren, J.; Cheynier, V. Structural determination of colourless and yellow dimers resulting from (+)-catechin coupling catalysed by grape polyphenoloxidase. *Phytochemistry*, **1996**, 42, 1279–1288.

Kritzinger, E. C.; Bauer, F. F.; Du Toit, W. J. Role of Glutathione in Winemaking: A Review. *Journal of Agricultural and Food Chemistry*, **2012**, 61, 269–277.

Ky, I.; Lorrain, B.; Jourdes, M.; Pasquier, G.; Fermaud, M.; Gény, L.; Rey, P.; Doneche, B.; Teissedre, P.-L. Assessment of grey mould (*Botrytis cinerea*) impact on phenolic and sensory quality of Bordeaux grapes, musts and wines for two consecutive vintages. *Australian Journal of Grape and Wine Research*, **2012**, 18, 215–226.

Li, H.; Guo, A.; Wang, H. Mechanisms of oxidative browning of wine. *Food Chemistry*, **2008**, 108, 1–13.

Lopes, P.; Richard, T.; Saucier, C.; Teissedre, P.-L.; Monti, J.-P.; Glories, Y. Anthocyanone A. A quinone methide derivative resulting from malvidin 3-*O*-glucoside degradation. *Journal of Agricultural and Food Chemistry*, **2007**, 55, 2698–2704.

Lorenzo, M.; Moldes, D.; Couto, S. R.; Sanromán, M. A.A. Inhibition of laccase activity from *Trametes versicolor* by heavy metals and organic compounds. *Chemosphere*, **2005**, 60, 1124–1128.

Macheix, J.-J.; Sapis, J.-C.; Fleuriet, A.; Lee, C. Y. Phenolic compounds and polyphenoloxidases in relation to browning in grapes and wine. *Critical Reviews in Food Science & Nutrition*, **1991**, 30, 441–486.

Maiorella, B.; Blanch, H. W.; Wilke, C. R. By-product inhibition effects on ethanolic fermentation by *Saccharomyces cerevisiae*. *Biotechnology and Bioengineering*, **1983**, 25, 103–121.

Marbach, I.; Harel, E.; Mayer, A. M. Molecular properties of extracellular *Botrytis cinerea* laccase. *Phytochemistry*, **1984**, 23, 2713–2717.

Ohsumi, Y.; Kitamoto, K.; Anraku, Y. Changes induced in the permeability barrier of the yeast plasma membrane by cupric ion. *Journal of Bacteriology*, **1988**, 170, 2676–2682.

Pedneault, K.; Provost, C. Fungus resistant grape varieties as a suitable alternative for organic wine production: Benefits, limits, and challenges. *Scientia Horticulturae*, **2016**, 208, 57–77.

Penninckx, M. A short review on the role of glutathione in the response of yeasts to nutritional, environmental, and oxidative stresses. *Enzyme and Microbial Technology*, **2000**, 26, 737–742.

Perino, A.; Vercesi, A.; Fregoni, M. Confronto tra diverse metodiche utilizzate nella determinazione dell'infezione da *Botrytis cinerea* sui mosti. *Vignevini*, **1994**, 7, 50–57.

Połeć-Pawlak, K.; Ruzik, R.; Lipiec, E. Investigation of Cd (II), Pb (II) and Cu (I) complexation by glutathione and its component amino acids by ESI-MS and size exclusion chromatography coupled to ICP-MS and ESI-MS. *Talanta*, **2007**, 72, 1564–1572.

Ribéreau-Gayon, P.; Dubourdieu, D.; Donèche, B.; Lonvaud, A. *Handbook of enology, The microbiology of wine and vinifications*, edition no. 2; John Wiley & Sons: West Sussex, UK, 2006. Vol. 1.

Salgues, M.; Cheynier, V.; Gunata, Z.; Wylde, R. Oxidation of grape juice 2-*S*-glutathionyl caffeoyl tartaric acid by *Botrytis cinerea* laccase and characterization of a new substance: 2, 5-di-*S*-glutathionyl caffeoyl tartaric acid. *Journal of Food Science*, **1986**, 51, 1191–1194.

Shanmuganathan, A.; Avery, S. V.; Willetts, S. A.; Houghton, J. E. Copper-induced oxidative stress in *Saccharomyces cerevisiae* targets enzymes of the glycolytic pathway. *FEBS Letters*, **2004**, 556, 253–259.

Sioumis, N.; Kallithraka, S.; Tsoutsouras, E.; Makris, D. P.; Kefalas, P. Browning development in white wines: Dependence on compositional parameters and impact on antioxidant characteristics. *European Food Research and Technology*, **2005**, 220, 326–330.

Smirnov, S. A.; Koroleva, O. V.; Gavrilova, V. P.; Belova, A. B.; Klyachko, N. L. Laccases from basidiomycetes. Physicochemical characteristics and substrate specificity towards methoxyphenolic compounds. *Biochemistry (Moscow)*, **2001**, 66, 774–779.

Ugliano, M. Enzymes in winemaking. In *Wine chemistry and biochemistry*; Moreno-Arribas, M. V.; Polo, M. C., Eds.; Springer: New York, USA, 2009, 103–126.

Ugliano, M.; Kwiatkowski, M.; Vidal, S.; Capone, D.; Siebert, T.; Dieval, J.-B.; Aagaard, O.; Waters, E. J. Evolution of 3-mercaptohexanol, hydrogen sulfide, and methyl mercaptan during bottle storage of Sauvignon blanc wines. Effect of glutathione, copper, oxygen exposure, and closure-derived oxygen. *Journal of Agricultural and Food Chemistry*, **2011**, 59, 2564–2572.

Vanni, A.; Anfossi, L.; Pessione, E.; Giovannoli, C. Catalytic and spectroscopic characterisation of a copper-substituted alcohol dehydrogenase from yeast. *International Journal of Biological Macromolecules*, **2002**, 30, 41–45.

Vest, K. E.; Zhu, X.; Cobine, P. A. Copper Disposition in Yeast. In *Clinical and Translational Perspectives on WILSON DISEASE*, edition no. 1; Kerkar, N.; Roberts, E. A., Eds.; Academic Press an imprint of Elsevier: Massachusetts, USA, 2019, 115–126.

Summary

Quality deterioration of wine still represents a great challenge in the wine industry. It may be attributed to several factors occurring at various stages of winemaking. One crucial aspect to ensure a high wine quality is to maintain metabolic activity and vitality of yeasts during fermentation. *Saccharomyces cerevisiae* is exposed to several stress conditions during winemaking, e.g., exposition to high copper levels. The presence of this heavy metal in must and wine emerges partially from the control of fungal infections of the grapevine with copper-based fungicides. This method has a long tradition in viticulture and is still applied today as one of a few instruments that are allowed in organic viticulture. If copper reaches levels above the essential amounts required by yeasts as a micronutrient, it leads to reduced enzyme activities up to total enzyme inhibition or a loss of membrane integrity. The application of copper-rich musts for vinification, thus, increases the risk of an impaired fermentation. The aim of this work was to analyze whether glutathione addition to the must influences copper-mediated stress during fermentation. Glutathione (GSH) is known to protect the yeast against heavy metals including induced oxidative stress in cells. The vitality and sugar metabolism of a yeast strain were distinctively affected at copper concentrations of 10–25 $mg \cdot L^{-1}$ in the must. In contrast, both parameters were almost unaffected at the same conditions for another copper-resistant strain. The addition of 20 $mg \cdot L^{-1}$ GSH positively influenced fermentation efficiency of the copper-sensitive strain by an increase in vitality, a shortening of the lag phase of sugar degradation together with an acceleration of the sugar degradation. This may reduce the risk of a stuck and sluggish fermentation. Additionally, a significant acetaldehyde accumulation was observed for the copper-sensitive strain under copper stress, which could be mitigated by GSH addition. An interaction of copper and GSH was proven by the detection of lower GSH concentrations in the must of microvinifications with higher copper concentrations. The ability of GSH to complex copper is part of the detoxification mechanism by GSH. Further proof for a copper-GSH-interaction was given by monitoring extra- and intracellular copper concentrations. GSH addition led to significant differences of both parameters, compared to the same microvinifications with copper exposition. It was further shown that copper reduced or inhibited specific alcohol dehydrogenase (ADH) activity in copper-stressed yeast cells. The activity was restored in fermentations with GSH at a copper concentration of 10 mg·L^{-1}. In microvinifications with low copper amounts (control and 2.5 mg·L^{-1}), GSH decreased the ADH activity of the copper-sensitive strain, presumably due to the complexation of essential copper amounts. The present work showed that GSH represents a useful tool to reduce copper-induced stress during fermentation.

Further wine quality deterioration may be attributed to the presence of laccase in must and wine as a consequence of processing *Botrytis cinerea*-infected grapes. The enzyme is secreted by the fungus to counteract the plant defense mechanisms. It is able to oxidize a wide range of phenolic compounds, produced as antifungal agents in the grapevine. A consequence of this laccase-catalyzed oxidation in must and wine is a discoloration in terms of browning in white wine and a color loss of red wine. The stability of the enzyme in wine and its resistance to commonly applied SO_2 amounts or heat treatments represent a huge problem in winemaking. The second aim of this work was to compare laccase activities in the secretomes of different *B. cinerea* isolates from red and white grape varieties. Distinct differences were observed in specific laccase activities of different isolates and their catalytic activities toward six typical wine phenols. Substrate half-life indicated fast oxidation of caffeic acid, ferulic acid, and malvidin 3-*O*-glucoside, and comparatively slower oxidation of (+)-catechin, gallic acid, and *p*-coumaric acid. Differences were observed between the laccases of different isolates by comparing half-life values for individual substrates. Identical product profiles for five out of six substrates were shown, but two strains isolated from Pinot noir and Portugieser grapes led to deviating product profiles during gallic acid oxidation. The impact of oxidation on the color was manifested in discoloration of reaction mixtures to yellow, greenish-yellow, or brown for (+)-catechin, gallic acid, and caffeic acid, respectively, while malvidin 3-*O*-glucoside oxidation led to a loss of the red color. The product formation rates, maxima in relative product concentrations as well as subsequent non-enzymatic decrease in product concentrations varied between the substrates and different laccases. The results revealed that laccases from different *B. cinerea* isolates exhibit different catalytic activities toward wine phenols and consequently distinctions in their impact on color deterioration of wine.

Zusammenfassung

Die Qualitätsminderung von Wein stellt nach wie vor eine große Herausforderung für die Weinindustrie dar. Sie kann auf viele Faktoren während verschiedener Stufen der Weinbereitung zurückgeführt werden. Ein wichtiger Aspekt für die Gewährleistung einer guten Weinqualität ist die Aufrechterhaltung der metabolischen Aktivität sowie der Vitalität von Hefen. *Saccharomyces cerevisiae* ist während der Gärung vielen Stressbedingungen ausgesetzt, beispielsweise einer Kupferexposition. Das Vorkommen dieses Schwermetalls in Most und Wein ist auf die Bekämpfung von Pilzinfektionen der Weinrebe mit kupferbasierten Fungiziden zurückzuführen. Diese Methode hat eine lange Tradition im Weinbau und findet auch heutzutage ihre Anwendung als eines der wenigen erlaubten Instrumente im biologischen Weinbau. Wenn der Kupfergehalt toxische Grenzen erreicht, die über den essentiellen Mengen des Mikronährstoffes für Hefen liegen, kann dieser zu reduzierten Enzymaktivitäten bis hin zu einer totalen Enzyminhibierung oder einem Verlust der Membranintegrität führen. Die Verwendung von kupferreichen Mosten in der Weinbereitung erhöht demzufolge das Risiko einer beeinträchtigten Gärung. Das Ziel dieser Arbeit war es zu untersuchen, ob eine Glutathionzugabe zu Most den kupferinduzierten Stress während der Gärung beeinflusst. Glutathion (GSH) ist dafür bekannt, Hefen gegen Schwermetalle und den induzierten oxidativen Stress zu schützen. Die Vitalität und der Zuckermetabolismus eines Hefestammes wurden bei einer Kupferkonzentration von 10–25 $mg \cdot L^{-1}$ im Most deutlich beeinträchtigt. Im Gegensatz dazu blieben beide Parameter unter denselben Bedingungen für einen weiteren, kupferresistenten Stamm beinahe unbeeinflusst. Die Zugabe von 20 $mg \cdot L^{-1}$ GSH bei Kupferstress beeinflusste die Fermentationseffizienz des kupfersensitiven Stammes durch eine Erhöhung der Vitalität, eine Verkürzung der lag-Phase des Zuckerabbaus zusammen mit einer Beschleunigung des Zuckerabbaus. Dies kann das Risiko einer verlangsamten und stockenden Gärung verringern. Zudem konnte eine signifikante Acetaldehydakkumulation für die kupfersensitive Hefe unter Kupferstress beobachtet werden, welche durch eine Glutathionzugabe abgeschwächt werden konnte. Eine Interaktion von Kupfer und Glutathion wurde durch den Nachweis von geringeren GSH-Konzentrationen im Most von Mikrovinifikationen mit höheren Kupferkonzentrationen belegt. Die Fähigkeit von GSH, Kupfer zu komplexieren, ist Teil des Detoxifikationsmechanismus durch GSH. Ein weiterer Hinweis für eine Kupfer-GSH-Interaktion wurde durch die Betrachtung der intra- und extrazellulären Kupferkonzentrationen gegeben. Eine GSH-Zugabe führte zu signifikanten Unterschieden bei beiden Parametern im Vergleich zu den gleichen Mikrovinifikationen mit alleiniger Kupferexposition. Weiterhin konnte gezeigt werden, dass Kupfer die spezifische Alkoholdehydrogenase (ADH)-Aktivität in kupfergestressten Hefezellen mindert oder inhibiert. Die Aktivität konnte durch GSH-Zugabe bei Kupferkonzentrationen von 10 $mg \cdot L^{-1}$ wieder hergestellt werden. In Mikrovinifikationen mit geringen Kupferkonzentrationen (Kontrolle und 2,5 $mg \cdot L^{-1}$) wurde die ADH-Aktivität des Kupfer-sensitiven Stammes durch GSH herabgesetzt, was vermutlich durch eine Komplexierung essentieller Mengen an Kupfer erklärt werden kann. Die vorliegende Arbeit zeigte, dass GSH ein nützliches Hilfsmittel darstellt, um kupferinduzierten Stress während der Gärung zu reduzieren.

Eine weitere Qualitätsminderung von Wein kann auf das Vorkommen von Laccase in Most und Wein zurückgeführt werden, als Ursache der Verarbeitung von *Botrytis cinerea*-infizierten Trauben. Das Enzym wird von dem Pilz abgegeben, um den pflanzlichen Abwehrmechanismen entgegenzuwirken. Es kann eine Vielzahl von phenolischen Verbindungen oxidieren, die als antifungale Stoffe von der Weinrebe produziert werden. Im Most und Wein hat diese laccasekatalysierte Oxidation Farbveränderungen in Form von Bräunungserscheinungen in Weißwein und Farbverlust von Rotwein zur Folge. Die Stabilität des Enzyms in Wein und seine Resistenz gegenüber standardmäßig eingesetzten SO_2-Mengen sowie Hitzebehandlungen stellen ein großes Problem in der Weinbereitung dar. Das Ziel dieser Arbeit war der Vergleich von Laccaseaktivitäten in den Sekretomen verschiedener *B. cinerea*-Isolate von roten und weißen Rebsorten. Es konnten deutliche Unterschiede in der spezifischen Laccaseaktivität der verschiedenen Isolate beobachtet werden sowie in ihrer katalytischen Aktivität gegenüber sechs typischen Weinphenolen. Die Substrat-Halbwertszeit zeigte eine schnelle Oxidation von Kaffeesäure, Ferulasäure und Malvidin-3-*O*-glucosid und eine vergleichsweise langsamere Oxidation von (+)-Catechin, Gallus- und *p*-Cumarsäure. Beim Vergleich der Halbwertszeiten für die einzelnen Substrate konnten Unterschiede für die Laccasen von verschiedenen Isolaten beobachtet werden. Es zeigten sich identische Produktprofile für fünf von sechs Substraten. Lediglich zwei Stämme, isoliert von Spätburgunder und Portugieser Trauben, führten bei der Oxidation von Gallussäure zu abweichenden Produktprofilen. Der Einfluss der Oxidation auf die Farbe zeigte sich in einer Farbveränderung der Reaktionsansätze zu gelb, grün-braun oder braun jeweils für (+)-Catechin, Gallus- und Kaffeesäure, während die Oxidation von Malvidin-3-*O*-glucosid zu einem Verlust der roten Farbe führte. Die Produktbildungsraten, die Maxima der relativen Produktkonzentrationen sowie die folgende nichtenzymatische Abnahme der Produktkonzentrationen variierten zwischen den Substraten und verschiedenen Laccasen. Die Ergebnisse machen deutlich, dass Laccasen von verschiedenen *B. cinerea*-Isolaten verschiedene katalytische Aktivitäten gegenüber Weinphenolen aufweisen und folglich einen unterschiedlichen Einfluss auf eine Farbbeeinträchtigung von Wein haben.

Acknowledgement

At this point, I would like to take the opportunity to thank all of the people who supported me during my time at the Institute of Nutritional and Food Sciences. Without all of them - and to only name a few - the successful development and completion of this thesis would not have been possible.

To start, I would like to thank Prof. Andreas Schieber for giving me the opportunity to work on the research project supported by the Ministry of Economics and Technology (via AiF) and the FEI (AiF 18645N). During my studies, he gave me the freedom and support to study interesting topics that arose for this dissertation.

I would also like to thank Prof. André Lipski for being my second supervisor as well as P.D. Christiane Dahl and Prof. Alf Lamprecht for their participation in the examination committee.

A huge thanks goes to Dr. Fabian Weber for supervising me during my doctoral studies. I appreciate his constant encouragement, support and that he always had an open door to discuss new ideas.

A lot of gratitude goes to all of my colleagues who enriched every day through pleasant teamwork, helpful conversations and outside of work activities such as fun and delicious cooking evenings, as well as many participating in different sports. I'm thankful for the time I spent with these colleagues, as many of them turned into valuable friendships that I will cherish from my time spent in Bonn. I would especially like to thank Rita who supervised me during my first practical course at the institute and, thus, she laid the foundation for my career at the institute. It was great to start the day with her early in the morning, where we had the opportunity to discuss new ideas, conduct experiments together, interpret results and, to celebrate those positive results together. Furthermore, a huge thanks to Lena, Michelle, Hannes, and Timo who always spent time with me at climbing or other sports after work. I would also like to thank Franziska for the collaboration on our project. It was great to have her support and to work through project meetings together. I also appreciate the one-on-one support, conversations and professionalism exuded with my continuous office colleague Lara, who shares the love of horses with me. Another good soul of the institute, Sandra, who never hesitated to help me in the lab and for that, I truly appreciate her. A big thanks to Lena and Nadine who offered me a home at the beginning and in the last months of my time in Bonn.

A special thank you goes to my students Julia and Lukas who contributed to the development of this thesis with a lot of diligence and great ideas. It was a pleasure to work in partnership with them.

I would also like to thank our project partners from the DLR in Neustadt (Weinstrasse) Prof. Dominik Durner, Dr. Pascal Wegmann-Herr, and Sebastian for interesting discussions and new ideas for the topics of this thesis.

Last but not least, a special thanks to my family for their support and love. Thank you mum, dad and Raphael. I'm very lucky to have you in my life, and truly appreciate you always being there to support me through my journey.

Curriculum Vitae

Sabrina Zimdars

Born 06 July 1990 in Marburg

Occupation

04/2015 – 06/2019	University of Bonn Institute of Nutritional and Food Sciences, Molecular Food Technology

Academic studies

04/2015 – 06/2019	University of Bonn PhD studies in Biology
10/2012 – 03/2015	University of Bonn Master of Science in Microbiology
10/2009 –10/2012	University of Koblenz Bachelor of Science in Environmental Sciences

School

08/2000 – 03/2009	Bertha-von-Suttner Gymnasium in Andernach

www.ingramcontent.com/pod-product-compliance
Ingram Content Group UK Ltd.
Pitfield, Milton Keynes, MK11 3LW, UK
UKHW021959190726
13853UKWH00004B/1629